PLUMBING

Design and Installation

Second Edition

Workbook

AMERICAN TECHNICAL PUBLISHERS, INC.
HOMEWOOD, ILLINOIS 60430-4600

ATP Staff

2 3 4 5 6 7 8 9 – 02 – 9 8 7 6 5 4 3

Printed in the United States of America

ISBN 0-8269-0616-8

Table of Contents

Introduction

Plumbing Design and Installation Workbook is designed to reinforce information presented in *Plumbing Design and Installation*. The textbook may be used as a reference to complete the learning activities in the workbook. Each chapter in the workbook covers information from the corresponding chapter in the textbook.

The question types used in the workbook include true-false, multiple choice, completion, matching, and short answer. For true-false questions, circle the T if the statement is true, or circle F if the statement is false. For multiple choice questions, write the letter of the correct answer in the answer blank next to the question. For completion questions, write the correct answer(s) in the answer blank next to the question. For matching questions, write the letter of the correct corresponding answer in the answer blank next to the question. For short answer questions, write the correct answer in the space provided.

In addition to questions, activities and projects are also included in each workbook chapter. The activities correlate with textbook chapter content and reinforce comprehension of related concepts and math principles. The projects provide additional opportunities for learners to research plumbing-related topics.

Information presented in *Plumbing Design and Installation* and *Plumbing Design and Installation Workbook* addresses common plumbing industry topics. To obtain information about all American Tech products, visit the American Tech web site at www.go2atp.com.

The Publisher

The Plumbing Trade

Chapter 1

Name_____ Date _____

True-False

T	**(F)** Boston	**1.** In 1652, Los Angeles developed the first city water system, which was primarily used for firefighting and domestic use.
(T)	F	**2.** An estimator is a person who estimates the cost of construction projects the contractor wants to bid on.
(T)	F	**3.** The duration of a plumbing apprenticeship is 5 years.
T	F	**4.** In 1829, the Tremont Hotel in Boston became the first hotel with indoor plumbing.
(T)	F	**5.** A superintendent is a journeyman who supervises a group of workers at a single location.
(T)	F	**6.** A plumbing apprentice is indentured to the JATC and is then assigned to a contractor.
T	**(F)**	**7.** The Cast Iron Soil Pipe Institute is a technical society dedicated to the improvement of the plumbing industry through the manufacturing, application, and installation of all types of drainage piping.
T	**(F)**	**8.** The Plastics Pipe Institute is the oldest trade organization in the construction industry.
(T)	F	**9.** The Standard Plumbing Code is used primarily in the Southeastern United States, including Alabama, Georgia, Tennessee, Mississippi, and Florida.
T	F	**10.** A detail drawing shows a small part of a floor plan, elevation, or section drawing at an enlarged scale.
(T)	F	**11.** The Plastic Pipe and Fittings Association represents manufacturers of plastic pipe, fittings, and solvent cements for plumbing and related applications.
(T)	F	**12.** The Plumbing Manufacturers Institute was originally called the Plumbing Brass Institute since it dealt solely with brass fitting manufacturers.
T	**(F)** Hoover	**13.** The Hunter Code was the forerunner of modern plumbing codes.
(T)	F	**14.** The Uniform Plumbing Code is used primarily in the Western United States, including California, Colorado, and Nevada.
T	F	**15.** Uppercase and lowercase letters may be used for abbreviations on prints.

1

T   F   **16.** The National Plumbing Code is used in Connecticut, Oklahoma, and Missouri.

T   F   **17.** A horizontal branch is a soil or waste pipe that receives only the discharge from fixtures on the same floor as the branch and extends horizontally from a stack.

T   F   **18.** A stop box is an adjustable cast iron box that is flush with the finish grade and is capped with a removable iron cover.

T   F   **19.** Schematic and isometric piping drawings are typically provided for small buildings such as residential dwellings.

T   F   **20.** A waste pipe ventilates the drainage system of a building and prevents trap siphonage and back pressure.

T   F   **21.** A water distribution pipe conveys water from the water service pipe to the point of use.

T   F   **22.** A building storm drain conveys rainwater and sewage.

T   F   **23.** A building drain branch is a soil or waste pipe that extends horizontally from the building drain and receives only the discharge from fixtures on the same floor as the branch.

T   F   **24.** The vent piping system provides circulation of air to or from a sanitary drainage system and provides air circulation within the sanitary drainage system to protect trap seals from siphonage or back pressure.

T   F   **25.** A sanitary drainage pipe removes wastewater and waterborne wastes from plumbing fixtures and conveys these wastes to the sanitary sewer or other point of disposal.

T   F   **26.** A building storm sewer conveys rainwater, but does not carry sewage.

T   F   **27.** Sewer gas is a mixture of vapors, odors, and gases found in sewers.

T   F   **28.** A sanitary sewer carries rainwater, surface water, groundwater, or similar nonpollutional wastes, but does not carry sewage.

Multiple Choice

_____   **1.** ___ water is water that is free from impurities which could cause disease or harmful physiological effects.
     A. Degassed
     B. Potable
     C. Soiled
     D. none of the above

_____   **2.** The first plumbing apprentice laws were passed in England early in the ___.
     A. 1400s
     B. 1500s
     C. 1600s
     D. 1700s

_____ **3.** A ___ is a person who is in charge of all field work for a contractor.
    A. supervisor
    B. general supervisor
    C. superintendent
    D. none of the above

_____ **4.** The first American patent for a plunger water closet was issued in ___ to William Campbell and James Henry.
    A. 1829
    B. 1857
    C. 1870
    D. 1876

_____ **5.** The Copper Development Association is a ___ that provides technical information regarding the proper and efficient use of copper and copper alloys in all applications, including plumbing, sprinkler systems, gas installations, and roofing.
    A. technical society
    B. trade association
    C. government department
    D. none of the above

_____ **6.** Currently, there are ___ states with their own occupational safety and health plans.
    A. 15
    B. 18
    C. 21
    D. 23

_____ **7.** Water closets were first patented in ___.
    A. England
    B. Italy
    C. France
    D. China

_____ **8.** In ___, William Smith invented the jet siphon water closet.
    A. 1829
    B. 1857
    C. 1870
    D. 1876

_____ **9.** The ___ is a technical society and primary developer of voluntary standards, related technical information, and services that promote public health and safety and contribute to the reliability of products, materials, and services.
    A. Mechanical Contractors Association of America
    B. Copper Development Association
    C. American Society for Testing and Materials
    D. American National Standards Institute

_____ **10.** A ___ is a building code developed by a regional committee of code enforcement officials.
   A. written specification
   B. model code
   C. variance code
   D. none of the above

_____ **11.** The ___ is a federal agency that requires all employers to provide a safe environment for their employees.
   A. Manufacturers Standardization Society of the Valve and Fitting Industry
   B. Copper Development Association
   C. American Society for Testing and Materials
   D. Occupational Safety and Health Administration

_____ **12.** ___ are written information from the architect and engineers that supplements the prints and provides additional details that cannot be shown on the prints or that require additional description.
   A. Specifications
   B. Model codes
   C. Details
   D. all of the above

_____ **13.** A(n) ___ is an orthographic view of a vertical surface without allowance for perspective.
   A. floor plan
   B. detail drawing
   C. elevation drawing
   D. section drawing

_____ **14.** The UA established the first nationally registered joint apprenticeship program in the United States in ___.
   A. 1912
   B. 1924
   C. 1936
   D. 1948

_____ **15.** The ___ is the original model plumbing code.
   A. Standard Plumbing Code
   B. National Plumbing Code
   C. International Plumbing Code
   D. A40 Standard Code

_____ **16.** Over ___ highly skilled UA members belong to approximately 420 local unions across North America.
   A. 160,000
   B. 210,000
   C. 290,000
   D. 450,000

_____ **17.** The ___ is a technical society organized for the development and improvement of the plumbing industry, national and international standards for valves, pipe fittings, flanges, pipe hangers, and related seals.
    A. Manufacturers Standardization Society
    B. American Society for Testing and Materials
    C. Plumbing Manufacturers Institute
    D. Mechanical Contractors Association of America

_____ **18.** ___ lines on a plot plan indicate the contour of a lot after excavation and grading.
    A. Dotted
    B. Dashed
    C. Solid
    D. Wavy

_____ **19.** A ___ cock is a valve placed on the water main to which the water service of a building is connected.
    A. gas
    B. curb
    C. corporation
    D. none of the above

_____ **20.** A ___ is any vertical line of soil, waste, or vent pipe extending through one or more stories.
    A. stack
    B. main
    C. branch
    D. riser

_____ **21.** A ___ cock is a valve installed on the water service to turn on or off the flow of potable water to a building.
    A. gas
    B. curb
    C. corporation
    D. none of the above

_____ **22.** A ___ is a water supply pipe connecting a fixture to the fixture branch pipe.
    A. water main
    B. riser
    C. fixture supply pipe
    D. building branch

_____ **23.** ___ lines on a plot plan indicate the existing contour of a building lot.
    A. Dotted
    B. Dashed
    C. Solid
    D. Wavy

_____ **24.** A(n) ___ piping drawing is a two-dimensional drawing of a piping system without regard to scale and exact location of the fixtures shown on the drawing.
    A. isometric
    B. schematic
    C. plan view
    D. none of the above

_____ **25.** A ___ is the part of the lowest piping of a drainage system that receives the discharge from soil, waste, and other drainage pipe inside the walls of the building and conveys it to the building sewer.
    A. fixture drain
    B. building drain
    C. building sewer
    D. waste stack

_____ **26.** A plumbing system constructed by a plumber to supply water to a building or remove the wastewater and waterborne waste material is the ___ system.
    A. sanitary drainage and vent piping
    B. potable water supply
    C. storm water drainage
    D. all of the above

_____ **27.** A(n) ___ piping drawing is a three-dimensional drawing of a plumbing drawing.
    A. isometric
    B. schematic
    C. plan view
    D. none of the above

_____ **28.** A ___ is a device used to measure the volume of water that passes through the water service.
    A. manometer
    B. water meter
    C. test gauge assembly
    D. barometer

_____ **29.** A ___ is any part of a plumbing system other than a riser, main, or stack.
    A. fixture
    B. casing
    C. waste pipe
    D. branch

_____ **30.** An estimator must be skilled in ___.
    A. mathematics
    B. printreading
    C. trade practices
    D. all of the above

_____ **31.** The ___ system includes the water service pipe, water distribution pipes, connecting pipes, fittings, control valves, and appurtenances inside or outside of a building but within the property lines.

    A. sanitary drainage and vent piping

    B. potable water supply

    C. storm water drainage

    D. all of the above

_____ **32.** A ___ stack is a vertical drainage pipe that extends one or more floors and receives the discharge of water closets, urinals, and similar fixtures.

    A. soil

    B. waste

    C. drainage

    D. none of the above

_____ **33.** A ___ is the part of a drainage system that extends from the end of a building drain and conveys its discharge to the public sewer, private sewer, individual sewage disposal system, or other point of disposal.

    A. fixture drain

    B. building drain

    C. building sewer

    D. waste stack

_____ **34.** A rainwater ___ conveys rainwater from the roof of a building to a storm drain or other point of disposal.

    A. riser

    B. leader

    C. drain

    D. stack

_____ **35.** A plumbing apprenticeship ___.

    A. is a minimum of five years

    B. includes approximately 8800 hours of work experience

    C. includes not less than 1000 hours of related instruction

    D. all of the above

Completion

_____ **1.** An individual who worked in the sanitary field in ancient Rome was called a(n) ___, taken from the Latin word "plumbum," which means lead.

_____ **2.** ___ is the trade that installs, removes, or modifies plumbing or drainage systems for the purpose of conveying a water supply and removing wastewater and waterborne waste.

_____ **3.** ___ was the first city in the world to adopt cast iron soil pipe for its water mains.

_____ 4. A(n) ___ is a person who has completed apprenticeship training and is skilled in a particular field.

_____ 5. In ___, John Randall Mann invented a siphonic water closet.

_____ 6. A(n) ___ is a person who enters into an agreement with an employer or JATC for a required period of time to receive instruction and learn a trade.

_____ 7. ANSI was founded in ___ by five engineering societies and three government agencies.

_____ 8. An individual or company responsible for the performance of construction work, including labor and materials, according to plans and specifications is a(n) ___.

_____ 9. OSHA regulations are included in Title ___ of the Code of Federal Regulations, Parts 1900–1999.

_____ 10. A(n) ___ is a plan view of a building that shows the arrangement of walls and partitions and provides information about windows, doors, and other features as they appear in an imaginary section taken approximately 5′ above floor level.

_____ 11. The Occupational Safety and Health Act of ___ requires all employers to provide work areas free from recognized hazards likely to cause serious harm.

_____ 12. A(n) ___ is a drawing showing a building structure and all of its components.

_____ 13. An interior ___ drawing is a scaled view that shows the shape and size of interior walls and partitions of a building.

_____ 14. A(n) ___ is a pipe that conveys potable water for public or community use from the municipal water supply source.

_____ 15. A(n) ___ drawing is a scaled view created by passing an imaginary cutting plane through a portion of a building.

_____ 16. A fixture ___ is a fitting or device that, when properly vented, provides a liquid seal to prevent the emission of sewer gases without affecting the flow of wastewater or waterborne waste.

_____ 17. A pipe extending from the corporation cock, water main, or other source of water supply to the water distribution system of a building is a(n) ___.

_____ 18. A fixture ___ is a water supply pipe that extends between a water distribution pipe and fixture supply pipe.

_____ 19. A(n) ___ conveys groundwater, rainwater, surface water, or similar nonpollutional wastes.

_____ 20. A(n) ___ is a vertical drainage pipe that extends one or more floors and receives the discharge of water closets, urinals, and similar fixtures.

_____ 21. The sanitary drainage system conveys wastewater and waterborne waste from the plumbing fixtures and appliances to the sanitary ___.

_____ **22.** A fixture ___ extends from a fixture trap to the junction of the next drainage pipe.

_____ **23.** A(n) ___ pipe conveys only liquid waste that is free from fecal matter.

_____ **24.** A roof drain receives rainwater collecting on a roof surface and discharges it into a rainwater ___.

_____ **25.** A(n) ___ cleanout allows access to the stacks for removal of stoppages from a pipe.

_____ **26.** ___ is any liquid waste containing animal or vegetable matter in suspension or solution and may include liquids containing chemicals in solution.

_____ **27.** A fitting with a removable cap or plug that is installed in a sanitary drainage pipe to allow access to the pipe for removing stoppages and for cleaning the interior of the pipe is a(n) ___.

_____ **28.** A(n) ___ cleanout is located near the front wall of a building where the building drain exits the building beginning at least 1′ outside the foundation footings.

Short Answer

1. Outline the history and development of plumbing in the United States, citing the major developments in plumbing technology.

_____

_____

_____

_____

_____

_____

2. Describe the curriculum and instruction of plumbing apprenticeship programs and list typical classroom and shop classes that may be required for an apprentice.

_____

_____

_____

_____

_____

Matching

Industry and Standards Organizations

_____ 1. Trade association

_____ 2. Government department

_____ 3. Technical society

A. Organization composed of groups of engineers and technical personnel united by a professional interest

B. Organization that represents the producers of specific products

C. Develops specifications such as the United States MIL STD

Plumbing Codes

_____ 1. SPC

_____ 2. UPC

_____ 3. NPC

_____ 4. IPC

A. Codeveloped by SBCCI, BOCA, and ICBO through the International Code Council

B. Developed by SBCCI and used primarily in the Southeastern United States

C. Developed by IAMPO and used primarily in the Western United States

D. Developed by BOCA and used in the Midwestern and Northeastern United States

Prints

_____ 1. Architectural prints

_____ 2. Structural prints

_____ 3. Mechanical prints

_____ 4. Electrical prints

_____ 5. Civil prints

A. Indicate the capacity of and placement of power plant systems, lighting, cable trays, conduit, wiring, and other electrical installations

B. Include information about overall placement of a building on the site, grading, elevations, and topographical information

C. Provide information about sizes, styles, and placement for foundations, beams, columns, joists, and other framing and load-bearing members

D. Include general building information, floor plans, elevations, section drawings, and detail drawings

E. Provide information about plumbing, heating, and air conditioning systems of a building

Vent Piping

_____ 1. Vent stack

_____ 2. Individual vent

_____ 3. Stack vent

_____ 4. Branch vent

A. Pipe that vents a single fixture trap

B. Vertical pipe that provides circulation of air to and from the drainage system

C. Extension of a soil or waste stack above the highest horizontal drain connected to the vent

D. Vent pipe connecting two or more individual vents with a stack vent or vent stack

Fixture and Appliance Symbols

_____ 1. Wall-hung urinal

_____ 2. Recessed bathtub

_____ 3. Tank-type water closet

_____ 4. Shower stall

_____ 5. Standard kitchen sink

_____ 6. Pedestal lavatory

_____ 7. Wall lavatory

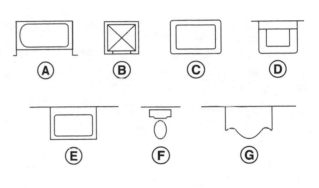

Pipe Fitting Symbols

_____ 1. 45° elbow

_____ 2. Tee

_____ 3. 90° elbow

_____ 4. Reducing elbow

_____ 5. Cross

Valve Symbols

_____ 1. Safety valve

_____ 2. Globe valve

_____ 3. Check valve

_____ 4. Gate valve

_____ 5. Angle globe valve

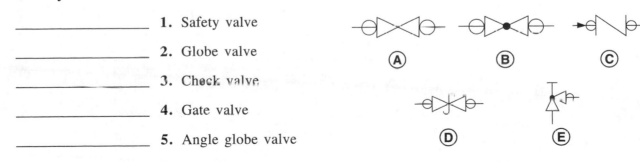

Potable Water Supply System

_____ **1.** Stop box

_____ **2.** Water service

_____ **3.** Fixture supply pipe

_____ **4.** Outside register

_____ **5.** Water meter

_____ **6.** Corporation cock

_____ **7.** Fixture branch

_____ **8.** Water distribution pipes

_____ **9.** Curb cock

_____ **10.** Water main

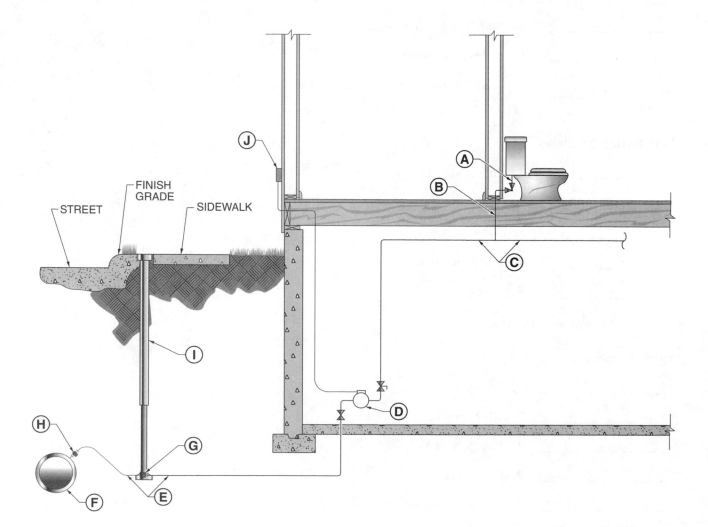

Sanitary Drainage and Vent Piping System

_____ **1.** Vent stack

_____ **2.** Building sewer

_____ **3.** Individual vent

_____ **4.** Building drain branch

_____ **5.** Branch vent

_____ **6.** Fixture trap

_____ **7.** Waste stack

_____ **8.** Sanitary sewer

_____ **9.** Roof jacket

_____ **10.** Front main cleanout

_____ **11.** Horizontal branch

_____ **12.** Soil stack

_____ **13.** Stack cleanouts

_____ **14.** Fixture drain

_____ **15.** Building drain

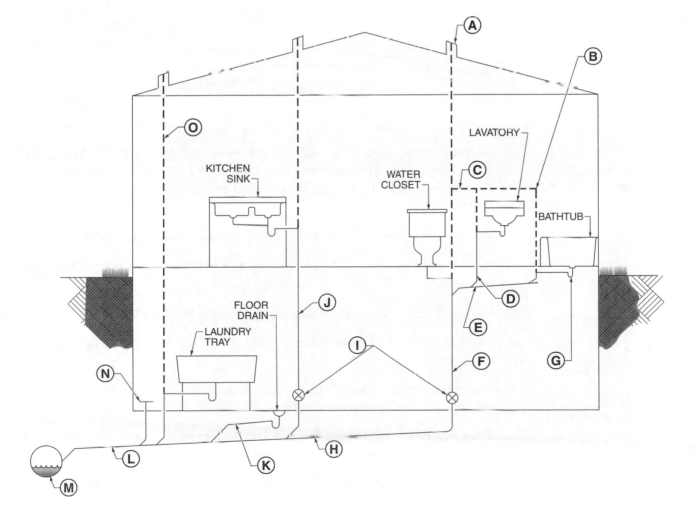

Activities

Interpreting Rough-in Sheets

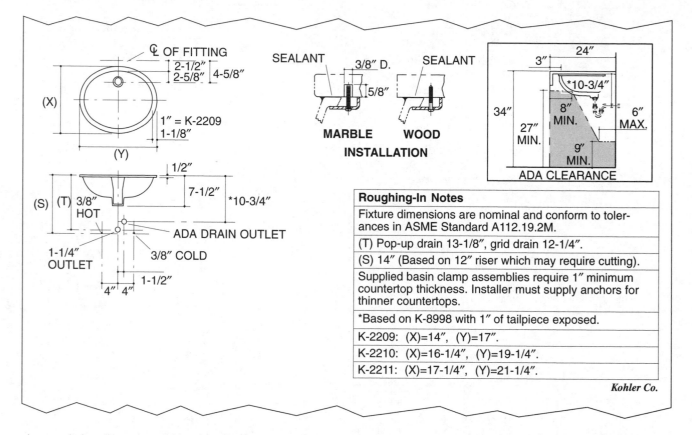

Kohler Co.

A rough-in sheet provides the dimensions needed to rough in the water supply and DWV piping for fixtures and appliances. Rough-in sheets typically include a dimensioned drawing of a fixture or appliance and notes to facilitate the installation. Using the information provided for the vitreous china undercounter lavatory, answer the following questions:

_____ 1. What are the dimensions of the lavatory bowl for model K-2209?

_____ 2. What is the thickness of the lavatory lip that attaches to the underside of the countertop?

_____ 3. Fixture dimensions conform to tolerances in ASME Standard ___.

_____ 4. The ADA-compliant drain outlet is located ___" to the right of the standard drain outlet.

_____ 5. The hot and cold water fixture supply pipes are located ___" to each side of the centerline of the lavatory fitting.

_____ 6. What size hole is required to be drilled on the underside of a marble countertop for attachment of the lavatory?

_____ 7. What are the dimensions of the lavatory bowl for model K-2211?

Sketching

1. Sketch an isometric piping drawing from the schematic piping drawing.

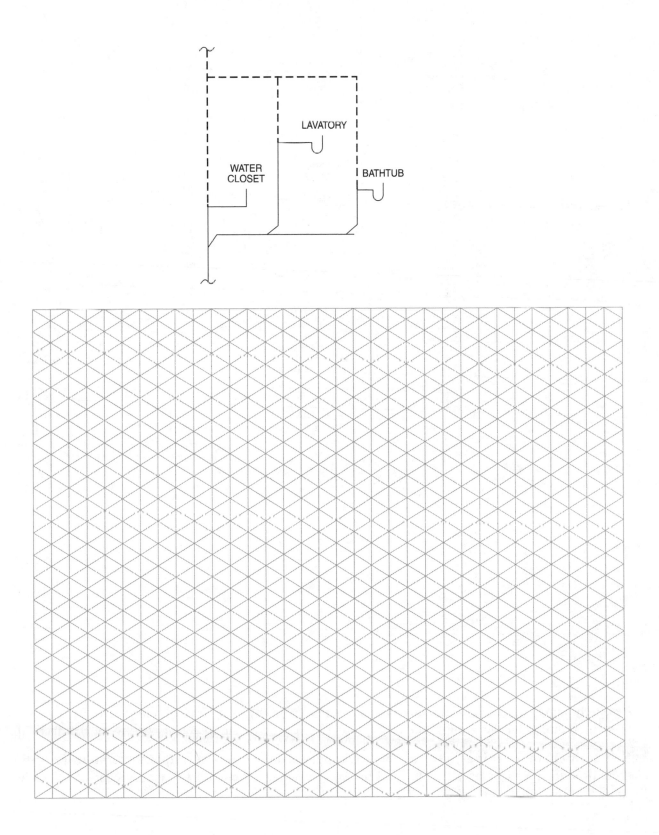

2. Sketch an isometric piping drawing from the schematic piping drawing.

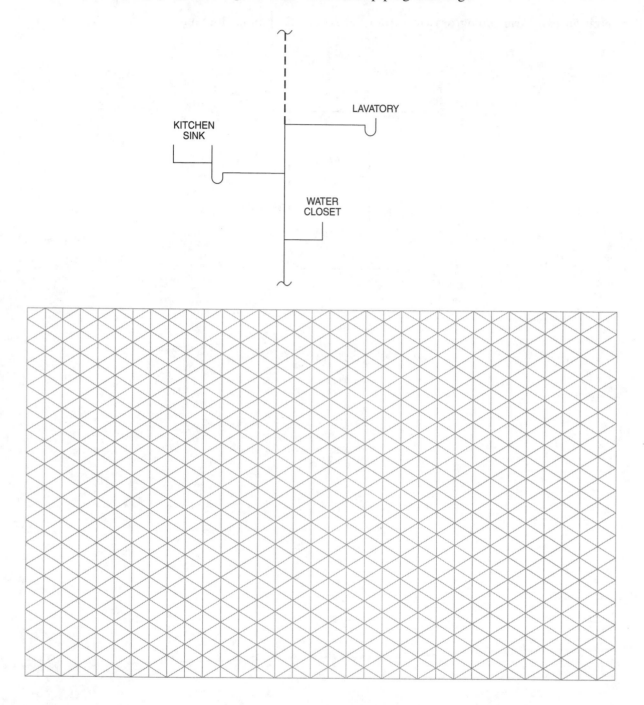

Projects

1. Research plumbing trade associations and technical societies by visiting their web sites or by requesting print literature from the groups. Identify and list each group's purpose and benefits of membership.

2. Contact plumbing fitting, fixture, and appliance manufacturers to request literature regarding their products. Visit their web sites and review the product literature to identify characteristics and applications of their products.

Name_____ Date _____

True-False

T   F   **1.** Clothing made of flammable synthetic materials may be worn on the job.

T   F   **2.** Goggles with tinted lenses protect against ultraviolet (UV) rays produced during soldering and brazing operations.

T   F   **3.** Employees must report accidents and safety hazards to the employer or supervisor.

T   F   **4.** Frames of safety glasses are designed to pop out the lenses if an impact occurs.

T   F   **5.** Metallic watches and rings should not be worn since serious injury may result if contact is made with an electrical circuit.

T   F   **6.** Hand tools such as rules or slip-joint pliers are always supplied by an employer.

T   F   **7.** The appropriate hand protection is determined by the duration, frequency, and degree of the hazard to hands.

T   F   **8.** Long objects such as pipe should be carried by two or more people.

T   F   **9.** Portable GFCIs commonly contain only one receptacle outlet protected by the module.

T   F   **10.** Most head injuries are the result of improper lifting procedures and could be prevented through proper planning and work procedures.

T   F   **11.** All ladders, regardless of the construction material, are manufactured to meet the same standards.

T   F   **12.** A trench box is a tool chest containing all of the supplies necessary for trench construction.

T   F   **13.** Oil-based paints may contain cadmium or chromium compounds as pigments, which could cause severe burns and may be fatal if swallowed or inhaled.

T   F   **14.** Single or extension ladders must always be raised with the ladder tip against the building.

T   F   **15.** The top support is the uppermost rung of the ladder

T   F   **16.** A GFCI protects against line-to-line contact hazards, such as a plumber holding two "hot" wires or a hot and neutral wire in each hand.

T F **17.** Rope slings are commonly used to lift medium to heavy loads.

T F **18.** A qualified person is a person who has special knowledge, training, and experience in performing specified duties.

T F **19.** OSHA regulations require that a scaffold be erected or dismantled only under the supervision of a qualified person.

T F **20.** At least two people are required to raise long extension ladders into position, with one person on each side of the ladder.

T F **21.** A safety sleeve is a moving element with a locking mechanism that is connected between a ladder carrier and a plumber harness or body belt.

T F **22.** Vertical lifelines must never have more than three people attached per line.

T F **23.** After a scaffold has been moved, the casters remain unlocked to allow movement while the scaffold is being used.

T F **24.** Confined spaces are isolated locations which protect workers from unsafe conditions such as oxygen deficiency, combustible gases, and/or toxic gases.

T F **25.** Lead is currently used to join bell-and-spigot cast iron soil pipe for sanitary waste and vent systems.

T F **26.** Kinking of slings and rope is prevented by hanging them from racks or rolling them onto spools.

T F **27.** Hazardous waste is recycled or blended for safe burning to recover its heat value.

T F **28.** A bloodborne pathogen is a part of the immune system which combats disease in the blood.

T F **29.** Workers must attend training prior to operating a material lift and have performance checked at least once every 10 years.

T F **30.** A hazardous material, such as cutting fluid, primer, and solvent cement, is a material capable of posing a risk to health, safety, and property.

T F **31.** Any type of fire extinguisher may be used on a fire to safely and quickly extinguish the fire.

T F **32.** Protective clothing and respirators are not necessary in sewer environments.

T F **33.** Material stored in a different container than originally supplied from the manufacturer must also be properly labeled.

T F **34.** If an MSDS is not provided, the employer must contact the manufacturer, distributor, or importer to obtain the missing MSDS.

T F **35.** Appropriate PPE, including a sealed full body suit with respirator, must be used when working in areas with airborne asbestos to avoid developing asbestosis.

T F **36.** Foam extinguishers or water must be used on electrical fires.

T   F   **37.** A ladder is a structure consisting of two side rails joined at intervals by steps or rungs for climbing up and down.

T   F   **38.** Trench boxes are used in stable or unstable soil.

Multiple Choice

_____ **1.** ___ gloves should be worn during manual threading and grinding operations.
  A. Leather
  B. Butyl
  C. Cotton
  D. Canvas

_____ **2.** To determine approximate noise reduction of ear protection devices in the field, ___ dB is subtracted from the noise reduction rating.
  A. 3
  B. 6
  C. 7
  D. 12

_____ **3.** An important feature of a protective helmet (hard hat) is that ___.
  A. it offers protection from falling or flying objects
  B. it is constructed of durable, lightweight material
  C. a shock-absorbing lining keeps the shell away from the head to provide ventilation
  D. all of the above

_____ **4.** The ___ is a federal agency that requires all employers to provide a safe environment for employees.
  A. American National Standards Institute
  B. Environmental Protection Agency
  C. Occupational Safety and Health Administration
  D. Mechanical Contractors Association of America

_____ **5.** ___ are an eye protection device with a flexible frame; they are secured on the face with an elastic headband.
  A. Contact lenses
  B. Face shields
  C. Safety glasses
  D. Goggles

_____ **6.** If injuries occur, medical assistance should be obtained ___.
  A. immediately
  B. at the end of a shift
  C. if the problem persists
  D. when convenient

_____ **7.** When operating a material lift, the ___ must be considered.
   A. type of surface and equipment and pedestrian traffic
   B. type of load and stability
   C. required load manipulation
   D. all of the above

_____ **8.** Electrical shock effects vary from a mild tingling sensation to ___.
   A. unconsciousness
   B. death
   C. a painful jolt
   D. none of the above

_____ **9.** Chemical hazards may enter the body through ___.
   A. inhalation
   B. absorption through the skin
   C. ingestion
   D. all of the above

_____ **10.** A(n) ___ is a folding ladder that stands independently of support.
   A. fixed ladder
   B. single ladder
   C. extension ladder
   D. stepladder

_____ **11.** When lifting heavy objects, it is important not to ___.
   A. bend at the waist, lifting an object using the shoulder muscles
   B. lift an object while keeping the back as straight as possible
   C. move forward after the whole body is vertical
   D. keep a load steady and close to the body

_____ **12.** Ladders over ___′ in length should be secured at the bottom.
   A. 10
   B. 12
   C. 15
   D. 18

_____ **13.** Heavy-duty, ___-wire extension cords must be used for tools and equipment.
   A. two
   B. three
   C. four
   D. six

_____ **14.** In cold climates, trenches extend ___ to prevent the water service from freezing.
   A. above the frost line
   B. below the foundation
   C. above the topsoil
   D. below the frost line

_____ **15.** A GFCI is rated to trip in as little as ___ of a second to prevent electrocution.
    A. ¹⁄₄₀
    B. ¹⁄₂₀
    C. ⅛
    D. ½

_____ **16.** A ___ is a rail secured to the top of uprights and erected along the exposed sides and ends of a platform.
    A. guardrail
    B. midrail
    C. toeboard
    D. none of the above

_____ **17.** Extension ladders are positioned on a 4:1 ratio or approximately a ___° angle of inclination.
    A. 60
    B. 75
    C. 80
    D. 85

_____ **18.** ___ is a common cause of trench failure.
    A. Instability of the trench bottom
    B. Surface water boiling up into the trench bottom
    C. Stress or deformation in the trench walls
    D. all of the above

_____ **19.** A ___ is a device that clamps securely to a rope and contains a ring to which a lifeline can be attached, providing protection from falls.
    A. lanyard
    B. harness
    C. rope grab
    D. body belt

_____ **20.** A ___ is a pivoting hook mechanism attached to the fly section of an extension ladder to hold the fly section at the desired height.
    A. pawl lock
    B. halyard
    C. center swivel pulley
    D. spreader

_____ **21.** A(n) ___ is an ear protection device made of moldable rubber, foam, or plastic, and inserted into the ear canal.
    A. earmuff
    B. earplug
    C. ear cuff
    D. hearing aid

_____**22.** The ___ is the lower section of an extension ladder.
  A. bed section
  B. working height
  C. fly section
  D. overlap

_____**23.** A confined space ___ is a document issued by an employer to allow and control entry into a confined space.
  A. waiver
  B. security pass
  C. entry permit
  D. none of the above

_____**24.** A safety net must be used where personnel are ___′ or more above ground, water, machinery, or any other solid surface when the worker is not otherwise protected by a body belt, lifeline, or scaffold.
  A. 15
  B. 25
  C. 30
  D. 50

_____**25.** A ___ is the degree of susceptibility of materials to burning based on the form or condition of the material and its surrounding environment.
  A. reactivity hazard
  B. flammability hazard
  C. specific hazard
  D. none of the above

_____**26.** ___ is securely connecting a body belt or harness directly or indirectly to an overhead anchor point.
  A. Anchoring
  B. Fastening
  C. Tying off
  D. Latching on

_____**27.** A symptom of occupational irritant contact dermatitis is ___.
  A. redness of the skin
  B. blisters or scales
  C. crusting of the skin
  D. all of the above

_____**28.** A motor vehicle is a ___ that can operate on a job site and on public roadways.
  A. light- or heavy-duty truck
  B. van
  C. piece of equipment
  D. all of the above

_____ **29.** ___ is securing equipment or materials in preparation for lifting by means of rope, cable, chain, or web sling.
- A. Rigging
- B. Loading
- C. Tying off
- D. Stabilizing

_____ **30.** ___ is classified as a hazardous material.
- A. Thread cutting oil
- B. Cleaning solvent
- C. Solvent cement
- D. all of the above

_____ **31.** ___ may be used as substitutes for protective clothing, especially when gloves or sleeves cannot be worn safely.
- A. Barrier creams
- B. Mitts
- C. Barrier shields
- D. all of the above

_____ **32.** The NFPA Hazard Signal System uses a ___-color diamond sign to display information about hazardous materials.
- A. two
- B. three
- C. four
- D. five

_____ **33.** A ___ is the track of a ladder safety system consisting of a flexible cable or rigid rail secured to the ladder or structure.
- A. safety sleeve
- B. harness
- C. carrier
- D. halyard

_____ **34.** ___ is a hazardous by-product of the decomposition of organic material found in a sewer.
- A. Sludge
- B. Soil
- C. Sewer gas
- D. Oxygen

_____ **35.** A ___ is a chemical agent that cleans and softens plastic pipe and fittings and allows solvent cement to penetrate more effectively into the pipe.
- A. cleaning solution
- B. solvent
- C. plastic pipe softener
- D. primer

_____ **36.** Fiberglass insulation may cause ___.
    A. headache, nausea, dizziness, and difficulty in breathing
    B. irritation and inflammation of the eyes
    C. digestive disturbances, weight loss, and general weakness
    D. all of the above

_____ **37.** ___ is (are) a lung disease(s) caused by inhaling dust containing crystalline silica particles.
    A. Emphysema
    B. Asbestosis
    C. Silicosis
    D. all of the above

_____ **38.** A(n) ___ is a document that details facts about an accident.
    A. accident report
    B. injury report
    C. accident record
    D. none of the above

_____ **39.** Class D fire extinguishers are identified by the color ___.
    A. blue
    B. yellow
    C. red inside
    D. green

Completion

_____ **1.** A(n) ___ is a unit of measure used to express the relative intensity of sound.

_____ **2.** Substance abuse policies prohibit employees from working while under the influence of illegal drugs, ___, or other controlled substances.

_____ **3.** ___ are an eye protection device with special impact-resistant glass or plastic lenses, reinforced frames, and side shields.

_____ **4.** Personal ___ equipment is safety equipment used to protect against safety hazards in the work area.

_____ **5.** A(n) ___ is an eye and face protection device that covers the entire face with a plastic shield, and is used for protection from flying objects or splashing liquids.

_____ **6.** Soiled protective clothing should be ___ regularly to reduce flammable hazards.

_____ **7.** A(n) ___ is a solid, liquid, or gas that exerts toxic effects by inhalation, absorption, or ingestion.

_____ **8.** ___ is the use of a portable protective device capable of withstanding cave-in forces.

_____ **9.** A(n) ___ is a condition that results when a body becomes part of an electrical circuit.

_____ **10.** Safety shoes with reinforced ___ toes protect against injuries caused by compression and impact.

_____ **11.** A(n) ___ is temporary wiring used to supply power to portable electric tools and equipment.

_____ **12.** A(n) ___ is a rubber, leather, or plastic pad strapped onto the knees for protection.

_____ **13.** An aerial ___ is a piece of extendable and/or articulating equipment designed to position personnel and/or materials in elevated locations.

_____ **14.** A ground fault ___ is a device that protects against electrical shock by detecting an imbalance of current in the normal conductor pathways and opening the circuit.

_____ **15.** The angle of ___ is the greatest angle above the horizontal plane at which a material will lay without sliding.

_____ **16.** Safe ladder climbing employs the ___-point contact method.

_____ **17.** ___ is the use of wood or metal members to temporarily support soil or construction materials.

_____ **18.** A(n) ___ scaffold is equipped with casters and is moved along a pipe run during installation.

_____ **19.** The ___ is the upper section of an extension ladder.

_____ **20.** ___ is the rigging, lifting, and transporting of a load by mechanical means.

_____ **21.** A(n) ___ is the likelihood of a material to cause, either directly or indirectly, temporary or permanent injury or incapacitation due to an acute exposure by contact, inhalation, or ingestion.

_____ **22.** A(n) ___ is a temporary or movable platform and structure for plumbers to stand on when working above floor level.

_____ **23.** ___ is the hoisting of equipment or materials by mechanical means.

_____ **24.** A(n) ___ is a rope or webbing that is attached to a worker and tie-off device to prevent the worker from hitting the ground or other object during a fall.

_____ **25.** A(n) ___ data sheet is printed documentation used to relay hazardous material information from the manufacturer, importer, or distributor to the worker.

_____ **26.** To produce ___ contact dermatitis, the irritant substance must infiltrate the outer layer of skin and come into direct contact with cells and tissue.

_____ **27.** A(n) ___ is a space large enough and configured so a plumber can enter and perform assigned work, has limited or restricted means for entry and exit, and is not designed for continuous occupancy.

_____ **28.** A(n) ___ hazard is the degree of likeliness of materials to release energy by themselves or by exposure to certain conditions or substances.

_____ **29.** A(n) ___ is a narrow excavation made below the surface of the ground.

_____ **30.** Occupational irritant ___ is an inflammation caused by irritants found on the job site that come into direct contact with the skin.

_____ **31.** A(n) ___ hazard is the extraordinary properties and hazards associated with a particular material.

_____ **32.** ___ is a mineral that has long, silky fibers in a crystal formation, and was a component of many building materials including fireproofing, pipe insulation, siding, and tile installed until the late 1980s.

_____ **33.** Areas that have been in contact with an irritant should be cleaned using a low-pH mild soap, ___, or neutralizers.

_____ **34.** A(n) ___ is a hazardous substance at a job site that can cause cancer.

_____ **35.** ___ is a chemical agent that penetrates and dissolves the surface of plastic pipe and fittings.

_____ **36.** Isocyanates are classified as ___ and are known to produce asthma and hypersensitivity pneumonitis.

Short Answer

1. Describe the proper lifting procedure.

_____

_____

_____

_____

_____

2. List five safety precautions when using power equipment.

_____

_____

_____

_____

_____

_____

3. List five safety precautions when using extension cords.

_____

_____

_____

_____

4. Describe the operation of a GFCI.

_____

_____

_____

_____

5. Distinguish between safety glasses, face shields, and goggles, and list an application for each.

_____

_____

_____

_____

_____

_____

6. List five safety precautions when working in and around trenches.

_____

_____

_____

_____

7. Describe the proper procedure for raising a ladder with the tip against a building.

_____

_____

_____

_____

_____

_____

8. List five safety precautions to observe when using ladders.

_____

_____

_____

_____

_____

9. Identify three causes of trench failure.

_____

_____

_____

10. List five safety precautions to observe when using an aerial lift.

_____

_____

_____

_____

_____

11. List five safety precautions to observe when working on or around scaffolds.

_____

_____

_____

_____

_____

_____

12. Describe the proper procedure for raising a ladder with the tip away from the building.

_____

_____

_____

_____

_____

_____

_____

_____

13. List five safety precautions when using hand tools.

_____

_____

_____

_____

_____

_____

14. List five safety precautions to observe when working around material lifts.

_____

_____

_____

_____

_____

15. Identify the nine sections of an MSDS and briefly describe each.

_____

_____

_____

_____

_____

_____

_____

_____

_____

_____

_____

_____

16. List four types of air-purifying respirators and identify an application for each.

_____

_____

_____

_____

_____

Matching

Shoring

_____ 1. Vertical shoring

_____ 2. Waler

_____ 3. Soldier pile

_____ 4. Lagging

_____ 5. Wood shoring

A. A vertical steel H-beam that is driven into the ground

B. Horizontal support member used to retain trench sheet piling

C. Shoring that uses wood components for stringers, braces, and piling

D. Shoring that uses opposing vertical structural members separated by screwjacks or hydraulic or pneumatic cylinders (cross braces)

E. Wood planks placed between steel soldier piles to retain earth on the side of a trench or excavation

Personal Protective Equipment

_____ 1. Gloves

_____ 2. Close-fitting trouser legs

_____ 3. Protective helmet

_____ 4. Safety shoes

_____ 5. Earplugs

_____ 6. Safety glasses

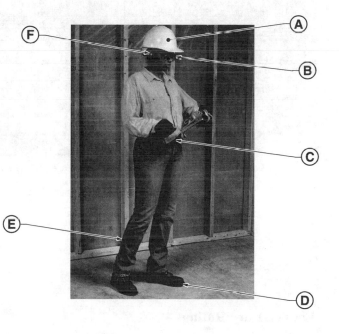

Soil Types

_____ **1.** Granular soil

_____ **2.** Cohesive soil

_____ **3.** Highly cohesive soil

A. Clay or soil with high clay content

B. Soil in which gravel or rock is held together with cohesive particles

C. Soil consisting of gravel, sand, or silts with little or no clay content

Aerial Lifts

_____ **1.** Articulating Z-boom

_____ **2.** Scissors

_____ **3.** Extendable S-boom

JLG Industries, Inc.

Ⓐ

Genie Industries

Ⓑ

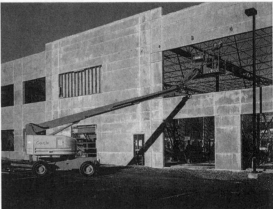

Genie Industries

Ⓒ

Ladder Duty Ratings

_____ **1.** Type IA

_____ **2.** Type I

_____ **3.** Type II

_____ **4.** Type III

A. Medium-duty, commercial, 225 lb capacity

B. Extra heavy-duty, industrial, 300 lb capacity

C. Light-duty, household, 200 lb capacity

D. Heavy-duty, industrial, 250 lb capacity

Freestanding Sectional Metal-Framed Scaffolds

_____ **1.** Footing base plate

_____ **2.** Cleat

_____ **3.** Midrail

_____ **4.** Guardrail

_____ **5.** Plank

_____ **6.** Cross brace

_____ **7.** Toeboard

_____ **8.** Hook-on ladder

_____ **9.** Coupling tube

_____ **10.** Bearers

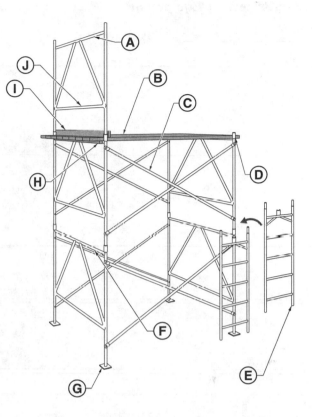

Required Eye Protection

_____ **1.** General duty or hammering

_____ **2.** General overhead work or sawing

_____ **3.** Soldering or brazing

_____ **4.** Chemical use

A. Goggles

B. Goggles and face shield

C. Tinted goggles

D. Safety glasses

Protective Helmets

_____ **1.** Class A

_____ **2.** Class B

_____ **3.** Class C

A. Protect against impact hazards; used in construction, manufacturing, and mining

B. Manufactured with lighter materials, yet provide adequate protection

C. Protect against high-voltage shock and burns, and impact or penetration from falling or flying objects

Fall-Protection Equipment

_____ **1.** Lanyard

_____ **2.** Lifeline

_____ **3.** Rope grab

_____ **4.** Safety net

_____ **5.** Harness

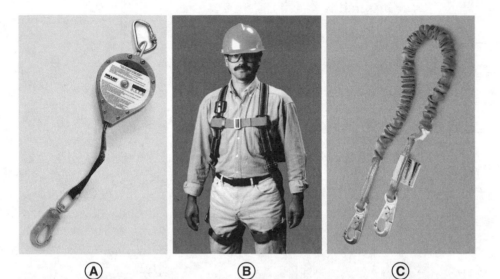

Ⓐ   Ⓑ   Ⓒ

Ⓓ   *The Sinco Group, Inc.*

Ⓔ   *Miller Equipment*

Manual Lifting Equipment

_____ **1.** Hand chain hoist

_____ **2.** Ratchet lever hoist

_____ **3.** Hand chain

_____ **4.** Hoist chain

A. Manually operated hoist used to raise and lower heavy sections of pipe or other loads by moving a ratchet lever back and forth

B. Continuous chain grasped by an operator to operate a hoist

C. Manually operated hoist used to raise and lower sections of pipe or other loads with a chain

D. Chain attached to the load

Fire Extinguisher Classification

_____ **1.** Electrical equipment

_____ **2.** Combustible metals

_____ **3.** Ordinary combustibles

_____ **4.** Commercial cooking grease

_____ **5.** Flammable liquids

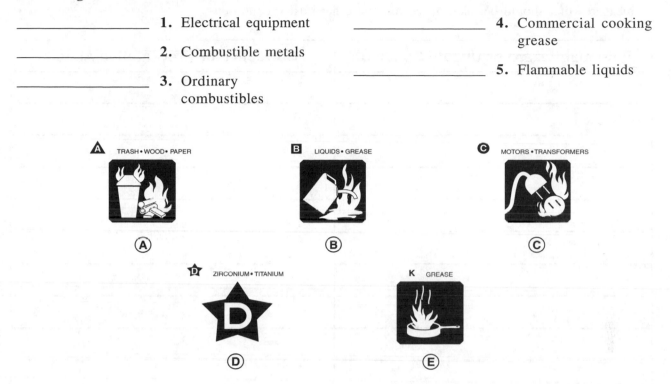

Activities

Safety Inventory

Conduct a safety inventory of your classroom or job site.

1. List the type and location of all safety equipment.

Safety Equipment Type	Location

2. List any potential hazards or unsafe conditions of tools and equipment. List corrective or preventative measures that should be taken to correct these hazards or conditions.

Potential Hazard or Unsafe Condition	Corrective or Preventative Action

Extension Ladder Placement

Extension ladders are positioned against a structure on a 4:1 ratio. For every 4′ of working height, 1′ of space is required at the base. Working height (WH) is the distance from the ground to the top support. The top support is the area of a ladder that makes contact with a structure. For example, an extension ladder with its top support at a working height of 12′ requires the base to be placed 3′ from the structure (12′ ÷ 4 = 3′). Determine the distance the base must be placed away from the structure when:

_____ **1.** Ladder base is placed ___′ from structure when WH = 20′.

_____ **2.** Ladder base is placed ___′ from structure when WH = 32′.

_____ **3.** Ladder base is placed ___′ from structure when WH = 10′.

_____ **4.** Ladder base is placed ___′ from structure when WH = 18′.

_____ **5.** Ladder base is placed ___′ from structure when WH = 26′.

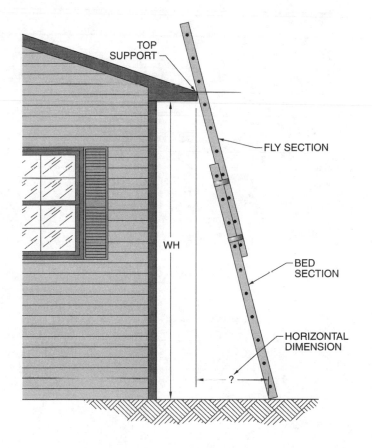

Projects

1. The electronic Compliance Assistance Tools (eCAT) developed by OSHA provide information regarding job site hazards and precautions that must be taken to avoid injury or death during construction operations. Review the Construction eCAT at www.osha-slc.gov/SLTC/construction_ecat and determine precautions that must be taken to avoid potential hazards associated with electricity, falls, being struck by moving objects, and trenching and excavation operations.

2. Obtain a catalog from a company that specializes in safety equipment and supplies, such as Lab Safety Supply, Inc., or visit their Web site. Identify PPE and other supplies and equipment that might be used during installation of plumbing systems.

Name_____ Date _____

True-False

T   F   **1.** A mixed number is a value that consists of a whole number and a proper fraction.

T   F   **2.** Dimensions are converted to mixed numbers by multiplying the whole number by the numerator and adding the value of the denominator.

T   F   **3.** When addition and multiplication calculations are completed, fractions often must be reduced to lowest terms.

T   F   **4.** When adding fractions, the denominators of the fractions must be the same prior to performing addition.

T   F   **5.** Fraction denominators must be the same prior to performing subtraction calculations.

T   F   **6.** When subtracting mixed numbers, the mixed numbers are converted to improper fractions prior to subtracting them.

T   F   **7.** When multiplying a fraction and a whole number, multiply the fraction denominator by the whole number.

T   F   **8.** A decimal can be considered a fraction whose denominator is 10, 100, 1000, or other multiple of ten.

T   F   **9.** When converting proper and improper fractions to their decimal inch equivalents, the numerator is divided by the denominator.

T   F   **10.** One millimeter is $\frac{1}{100}$ meter.

T   F   **11.** One gram is $\frac{1}{100}$ kilogram.

T   F   **12.** When converting from a larger unit such as a meter to a smaller unit such as a centimeter, the decimal point is moved to the right.

T   F   **13.** When converting from a smaller unit such as a meter to a larger unit such as a kilometer, the decimal point is moved to the left.

T   F   **14.** The offset distance is the vertical distance between the centerlines of the two horizontal sections of pipe to be offset.

T   F   **15.** The center to center distance is the theoretical length of the travel piece.

T   F   **16.** Rise is the vertical distance the pipe is offset.

T   F   **17.** Liquids have a fixed volume, but their shape is determined by the shape of the container holding them.

T   F   **18.** The volume of a gas is determined by the shape of the container holding it.

T   F   **19.** When water freezes, it expands by approximately $8\frac{1}{3}$ % of its original volume, but its weight does not change.

T   F   **20.** A square or rectangular surface has two dimensions, typically length and width, which are multiplied together to determine the area.

T   F   **21.** The capacity of water heaters and water storage tanks is typically expressed in gallons.

T   F   **22.** When draining a water pipe in a building, the capacity of the pipe should be known so that adequate-size containers are obtained.

T   F   **23.** Regardless of the pipe material, the pressure of water in pipe should be known before removing pipe hangers and other supports.

T   F   **24.** When converting decimal inch values to fractional inch equivalents, divide the decimal inch value by the numerator of a fraction that has the same numerator and denominator.

Multiple Choice

_____ **1.** A(n) ___ fraction is a fraction with a numerator smaller than its denominator.
  A. proper
  B. whole number
  C. improper
  D. calculating

_____ **2.** When performing addition or subtraction calculations involving proper and improper fractions, the fraction ___ must be the same.
  A. dimensions
  B. remainder
  C. numerators
  D. denominators

_____ **3.** Field conversion assumes that each ⅛″ equals ___′ and each full 1″ equals ___.
  A. .01; .08
  B. .1; .8
  C. .75; .25
  D. .8; .1

_____ **4.** The decimal equivalent of ⅛″ is ___.
A. .0625
B. .125
C. .25
D. .875

_____ **5.** Decimal foot values are converted to their inch equivalents by multiplying the decimal foot value by ___.
A. 10
B. 11
C. 12
D. 13

_____ **6.** The SI metric system is a decimal measurement system based on the ___.
A. meter and kilogram
B. meter and kilometer
C. meter and centimeter
D. kilogram and centimeter

_____ **7.** One meter is equal to ___′.
A. 1.8
B. 2.2
C. 3.28
D. 4.1

_____ **8.** One gram is equal to ___ oz.
A. .001
B. .025
C. .035
D. .5

_____ **9.** A percent grade is the fall (in feet) per ___′ of pipe run.
A. 25
B. 50
C. 75
D. 100

_____ **10.** Water obtains its maximum density at ___°F.
A. 30.1
B. 32
C. 38.1
D. 39.1

_____ **11.** The ___ length is the actual length of the travel piece.
A. end-to-end
B. fitting
C. constant
D. back-to-back

_____ **12.** The ___ is the distance from the center of one offset fitting to the center of the other offset fitting.
      A. travel width
      B. travel length
      C. advance
      D. offset distance

_____ **13.** The ___ is the measurement from the end of the travel piece when it is properly installed to the center of a fitting.
      A. uniform spread
      B. fitting angle
      C. center allowance
      D. fitting allowance

_____ **14.** A(n) ___ is an offset in which the centerline of the pipe changes direction both horizontally and vertically.
      A. horizontal/vertical offset
      B. offset distance
      C. rolling offset
      D. centerline offset

_____ **15.** The ___ is the distance the drainage pipe drops in its given length, and is expressed in inches.
      A. cover
      B. fall
      C. run
      D. pitch

_____ **16.** Water expands ___% as its temperature rises from 32°F to 212°F.
      A. 1½
      B. 2½
      C. 3½
      D. 4½

_____ **17.** ___ is the amount of water pressure in a column between points at different elevations, and is expressed in feet.
      A. Cubic feet
      B. Head
      C. Volume
      D. Foot hold

_____ **18.** ___ is the surface measurement within two boundaries, and is expressed in square units such as square feet.
      A. Area
      B. Base
      C. Constant
      D. Spread

_____ **19.** A 1′ water column exerts ___ psi on its base.
   A. .134
   B. .144
   C. .344
   D. .434

_____ **20.** ___ is the three-dimensional capacity of an object such as a tank or pipe, and is expressed in cubic units such as cubic inches or cubic feet.
   A. Dimension capacity
   B. Diameter
   C. Cubic capacity
   D. Volume

Completion

_____ **1.** A(n) ___ is a fraction with a numerator larger than its denominator.

_____ **2.** When an addition or subtraction calculation results in an improper fraction, the improper fraction is converted to a ___.

_____ **3.** Prior to subtracting mixed numbers, the mixed numbers are converted to ___ fractions.

_____ **4.** Division calculations involve dividing the ___ by a divisor to obtain a quotient.

_____ **5.** The memory method of conversion is referred to as ___ conversion.

_____ **6.** Metric prefixes indicating values less than 1 end in the letter ___.

_____ **7.** The ___ is the basic unit of weight in the SI metric system.

_____ **8.** The ___ is the vertical distance between the centers of two offset fittings.

_____ **9.** The ___ is the basic unit of length in the SI metric system.

_____ **10.** A(n) ___ is a combination of elbows or bends that brings one section of the pipe out of line but into a line parallel with another section.

_____ **11.** ___ is the horizontal center-to-center distance between pipes in a multiple-pipe offset.

_____ **12.** One ___ is ¹⁄₁₀₀ gram.

_____ **13.** ___ is the slope of a horizontal run of pipe and is expressed as a fractional inch per foot length of pipe.

_____ **14.** One cubic foot of water weighs approximately ___ lb.

_____ **15.** If a pipe that is full of water freezes, the expansion of the water when it turns to ice exerts a pressure of approximately ___ psi on the inside of the pipe.

_____ **16.** ___ water is water under pressure that is heated above 212°F without becoming steam.

_____ **17.** The ___ is the distance across the circumference of a circular surface passing through the center point.

_____ **18.** The ___ is the distance from the center point to the circumference of a circular surface, and is one-half of the diameter.

_____ **19.** When an architectural unit includes a(n) ___ fraction, the fraction is first converted to a decimal inch equivalent.

_____ **20.** One of the main advantages of the SI metric system is that one metric unit is converted to another unit by multiplying or dividing by multiples of ___, which moves the decimal point to the left or right.

Matching

Decimal Equivalents

_____ **1.** $\frac{1}{16}$ **A.** .25

_____ **2.** $\frac{1}{8}$ **B.** .5625

_____ **3.** $\frac{3}{16}$ **C.** .875

_____ **4.** $\frac{1}{4}$ **D.** .1875

_____ **5.** $\frac{3}{8}$ **E.** .625

_____ **6.** $\frac{1}{2}$ **F.** .0625

_____ **7.** $\frac{9}{16}$ **G.** .5

_____ **8.** $\frac{5}{8}$ **H.** .75

_____ **9.** $\frac{3}{4}$ **I.** .125

_____ **10.** $\frac{7}{8}$ **J.** .375

Metric Prefixes

_____ **1.** Centi-   **A.** Thousands

_____ **2.** Deka-   **B.** Tens

_____ **3.** Kilo-   **C.** Tenths

_____ **4.** Hecto-   **D.** Hundreds

_____ **5.** Milli-   **E.** Hundredths

_____ **6.** Deci-   **F.** Thousandths

Short Answer

1. Describe the procedure for calculating a multiple-pipe offset with uniform spreads.

_____

_____

_____

_____

_____

2. Explain why ice floats in water.

_____

_____

_____

_____

_____

Activities

Working with Whole Numbers, Fractions, and Mixed Numbers

_____ **1.** $\frac{1}{2} + \frac{1}{4}$

_____ **2.** $\frac{3}{4} + \frac{1}{2}$

_____ **3.** $2\frac{1}{8} - \frac{1}{4}$

_____ **4.** $3\frac{1}{2} - \frac{7}{8}$

_____ **5.** $\frac{1}{4} \times 3$

_____ **6.** $2\frac{1}{2} \div 2$

_____ **7.** $1\frac{1}{8} + 2\frac{1}{4}$

_____ **8.** $4 \times 2\frac{1}{2}$

_____ **9.** $\frac{1}{2} \div 2$

_____ **10.** $4\frac{1}{2} - \frac{5}{8}$

Working with Decimals

_____ **1.** 4.56 + 2.125

_____ **2.** 5.25 + 1.75

_____ **3.** 10.5 − 2.45

_____ **4.** 6.875 − 1.5

_____ **5.** 4 × .125

_____ **6.** 6.5 × 3.25

_____ **7.** 10 ÷ 2.5

_____ **8.** 12.375 ÷ 2

_____ **9.** 14.25 − 6.875

_____ **10.** 12.375 − 1.25

English and SI Metric Conversions

_____ **1.** 4″ = ___ cm

_____ **2.** 100 sq ft = ___ m^2

_____ **3.** 1000 cm^3 = ___ cu in.

_____ **4.** 4 kg = ___ lb

_____ **5.** 128 m = ___ ′

_____ **6.** 40 oz = ___ g

_____ **7.** 1 gal. = ___ l

_____ **8.** 35 cu ft = ___ m^3

_____ **9.** 2 l = ___ gal.

_____ **10.** 50.9 cm = ___ ″

Piping Calculations

_____ **1.** Determine the E-E length of a piece of 1½″ copper tube if the offset distance is 18″ and the offset is made with 1½″ 45° copper elbows (FΛ = 1.09″).

_____ **2.** Determine the E-E length of a piece of 4″ ABS DWV pipe if the offset distance is 6″ and the offset is made with 4″ 60° elbows (FA = 1.625″).

_____ **3.** Determine the difference in length and C-C distance of three pipes running 10″ C-C, and which are offset 16″ with 45° elbows.

_____ **4.** Determine the difference in length and C-C distance of five pipes running 8″ C-C, and which are offset 12″ with 60° elbows.

_____ **5.** Determine the travel length in a rolling offset made with a 45° wye and 60° elbow when the rise is 17½″.

_____ **6.** Determine the travel length for a 45° rolling offset with an 8″ rise and 14″ spread.

Working with Water

_____ **1.** 1 cu ft of water = ___ lb

_____ **2.** 1 gal. of water = ___ cu in.

_____ **3.** 1 gal. of water = ___ lb

_____ **4.** 140′ of head = ___ psi

_____ **5.** 85′ of head = ___ psi

_____ **6.** 22.5 psi = ___′ of head

_____ **7.** 50 psi = ___′ of head

Area and Volume Calculations

_____   **1.** Determine the area of a 16″ diameter circle.

_____   **2.** Determine the volume of a rectangular tank measuring 2′ × 4′ × 8′.

_____   **3.** Determine the volume of a cylindrical tank measuring 2′ diameter by 6′ tall.

_____   **4.** Determine capacity of a cylindrical tank measuring 18″ diameter by 12′ tall.

_____   **5.** Determine the weight of water in a section of pipe measuring 6″ diameter by 12′ long.

_____   **6.** Determine the weight of water in a section of pipe measuring 10″ diameter by 15′ long.

Projects

1. The solid state of water is ice and the gaseous state of water is steam. Water, like all matter, consumes a given amount of space (volume) and has weight. The volume of water expands when it is heated or cooled from 39.1°F, but its weight may or may not change. Find studies that have been performed on water and other substances and substantiate the studies' volume and weight findings.

2. Obtain short (6″–24″) sections of several types of pipe including ABS, PVC, copper, and cast iron soil pipe. Cap or plug one end of each section of pipe, weigh it, and record the weights. Fill the pipes with water, weigh them again, and record their weights. Determine the volume of water in each pipe given the inside diameter and length of the pipe. Compare the water weight and volume for each section of pipe.

Name_____ Date _____

True-False

T   F   1. "Size" is the general term commonly used when referring to pipe diameter, and should not be taken as the actual outside diameter of the pipe.

T   F   2. An expansion loop provides an area for plastic tubing to expand and contract without stressing.

T   F   3. Plastic fittings are manufactured by the extrusion process in which plastic resins are heated, softened, and forced through a die.

T   F   4. PVC pipe and fittings are used for sanitary drainage and vent piping, aboveground and underground storm water drainage, water mains, and water service lines.

T   F   5. PVC Schedule 40 pipe and fittings are used for industrial pressure applications.

T   F   6. Type K copper tube has the thickest wall, followed by type L, type M, and type MWV with the thinnest wall.

T   F   7. PEX tubing manufactured using the silane or radiation process has thermal memory.

T   F   8. ABS pipe and fittings are used for water supply piping.

T   F   9. It is acceptable to interchange PVC and ABS pipe and fittings in piping systems.

T   F   10. Crimping rings are not required for PEX tubing manufactured with the Engel process.

T   F   11. Since the late 1940s, over 14 billion pounds of copper tube has been manufactured in the United States for plumbing and plumbing-related applications.

T   F   12. DWV fittings have deeper solder sockets than pressure fittings.

T   F   13. Rolled groove joint copper fittings are used for aboveground potable water supply applications.

T   F   14. No-hub and bell-and-spigot cast iron soil pipe and fittings are primarily used for water supply piping.

T   F   15. Flared joint fittings are used with drawn copper tube.

T   F   16. Drawn copper tube is also referred to as hard copper.

T   F   17. Nominal pipe size is the approximate inside diameter of steel pipe.

T   F   **18.** Drawn copper tube is available in 40′–100′ coils.

T   F   **19.** Class 300 malleable iron fittings are standard weight fittings.

T   F   **20.** Galvanized grooved joint fittings are usually used for water piping.

T   F   **21.** Annealed copper tube is available in 20′ straight lengths and in coils ranging from 40′ to 100′ coils.

T   F   **22.** Wrought copper fittings are made from 99.9% pure copper.

T   F   **23.** Change-in-direction DWV fittings have a larger radius than pressure fittings to prevent stoppage.

T   F   **24.** T and C steel pipe has threads on both ends of the pipe and a coupling on one end.

T   F   **25.** The pressure rating for a rated valve is marked in raised letters on the inlet of the valve.

T   F   **26.** Globe valves must be installed with the flow direction arrow pointing in the downstream direction.

T   F   **27.** A frost-free sillcock is installed with a slight pitch toward the interior of the building.

T   F   **28.** When the wedge-shaped disk is retracted from the seat, a gate valve permits a straight and unrestricted fluid flow through the valve.

T   F   **29.** Opening or closing a core cock requires a 360° rotation of the valve handle.

T   F   **30.** When identifying reducing fittings, the sizes of the inlet and outlet of the run are listed first, followed by the size of other outlets.

T   F   **31.** Valve bodies for 2½″ and larger valves are typically manufactured from cast bronze with bronze internal components.

T   F   **32.** Split-wedge disk gate valves should be installed with the valve stem in the horizontal position.

T   F   **33.** Globe valves are recommended on installations requiring frequent operation, throttling, and/or a positive shutoff when closed.

T   F   **34.** A ball valve requires a 90° rotation of the handle to open or close the valve.

T   F   **35.** Globe valves have nonrising stems.

T   F   **36.** A pressure relief valve automatically lowers excessive pressure in a closed plumbing system.

T   F   **37.** A backwater valve is used to prevent the backflow of water in water supply piping.

T   F   **38.** Adequate clearance must be provided above a rising stem-inside screw gate valve since the stem and handwheel rise during operation.

T   F   **39.** A relief valve opens when pressure and/or temperature in a closed plumbing system exceeds safe operating limits.

T   F   **40.** A globe valve permits unrestricted fluid flow through the valve.

T   F   **41.** Pressure-reducing valves are installed near the water meter to reduce excessive water main pressure.

Multiple Choice

_____   **1.** Pipe is a cylindrical tube used for conveying ___ from one location to another.
     A. wastewater
     B. potable water
     C. waterborne waste
     D. all of the above

_____   **2.** Most standards require plastic pipe to be marked with ___.
     A. manufacturer name or trademark
     B. pipe size
     C. resin type
     D. all of the above

_____   **3.** ___ is a thermosetting plastic made from medium- or high-density cross-linkable polyethylene and is used for water-service piping and cold and hot water distribution piping.
     A. PE
     B. CPVC
     C. ABS
     D. PEX

_____   **4.** ___ are a family of synthetic materials manufactured from petroleum-based products and chemicals.
     A. Plastics
     B. Neoprene
     C. Resins
     D. None of the above

_____   **5.** Plastic pipe is manufactured by the ___ process in which plastic resins are heated, softened, and forced through a cylindrical die.
     A. injection molding
     B. extrusion
     C. thermoforming
     D. gleaning

_____   **6.** ABS pipe and fittings have an operational temperature range of ___°F.
     A. 0 to 100
     B. –20 to 140
     C. –40 to 180
     D. –60 to 240

_____ **7.** Copper tube is used for ___.
    A. fire suppression systems
    B. water supply and distribution piping
    C. HVAC applications
    D. all of the above

_____ **8.** A ___ is a plumbing design in which centrally located manifolds distribute water to each fixture with dedicated hot and cold water lines.
    A. home run
    B. manifold system
    C. direct line
    D. main system

_____ **9.** CPVC are ___-colored thermoplastic materials specially formulated to withstand higher temperatures than other plastics.
    A. black
    B. cream
    C. white
    D. orange

_____ **10.** PEX offers many advantages over other plumbing materials, such as copper, including ___.
    A. faster installation
    B. high-temperature and high-pressure resistance
    C. chemical and corrosion resistance
    D. all of the above

_____ **11.** The ___ process is a PEX manufacturing process in which polyethylene is subjected to high-energy electrons to form the cross-linked bond.
    A. silane
    B. radiation
    C. Engel
    D. all of the above

_____ **12.** Drawn and annealed copper tube is stamped every ___″ with the tube type, manufacturer name or trademark, and country of origin.
    A. 6
    B. 12
    C. 18
    D. 24

_____ **13.** The ___ process is a PEX manufacturing process in which peroxides release molecules for cross-linking.
    A. silane
    B. radiation
    C. Engel
    D. all of the above

_____ **14.** Flared joint fittings are used with type ___ annealed copper tube.
    A. K
    B. L
    C. M
    D. all of the above

_____ **15.** No-hub cast iron soil pipe and fittings are available in sizes ranging from ___".
    A. ½ to 12
    B. 1½ to 15
    C. 2½ to 18
    D. 4 to 24

_____ **16.** Schedule ___ steel pipe is standard weight steel pipe.
    A. 40
    B. 60
    C. 80
    D. 120

_____ **17.** Cast iron soil pipe and fittings are ___.
    A. nonabsorbent
    B. corrosion-resistant
    C. leakproof
    D. all of the above

_____ **18.** CPVC pipe is joined by ___.
    A. threading
    B. solvent cementing
    C. compression fittings
    D. flared fittings

_____ **19.** Cast copper alloy fittings are an alloy of ___ and are cast in sand molds.
    A. copper
    B. tin
    C. zinc
    D. all of the above

_____ **20.** Compression joint fittings are used with type ___ annealed copper tube.
    A. K
    B. L
    C. M
    D. all of the above

_____ **21.** Steel pipe and fittings are used for ___ piping systems.
    A. sanitary waste and vent
    B. water distribution
    C. storm water drainage
    D. all of the above

_____ **22.** A ___ nipple is threaded on both ends and has a short portion of unthreaded pipe in the middle.
  A. shoulder
  B. close
  C. wide
  D. no-shoulder

_____ **23.** Bell-and-spigot cast iron soil pipe and fittings designated as service weight are available in sizes ranging from ___″.
  A. ½ to 12
  B. 2 to 15
  C. 2½ to 18
  D. 4 to 24

_____ **24.** A ___ fitting is a pipe fitting in which all openings are the same dimension.
  A. reducing
  B. straight
  C. standard
  D. none of the above

_____ **25.** Gate valves are available in ___ configuration.
  A. nonrising stem-inside screw
  B. rising stem-inside screw
  C. rising stem-outside stem and yoke
  D. all of the above

_____ **26.** A ___ valve is designed to be used in its fully open or fully closed position.
  A. control
  B. throttling
  C. gate
  D. rated

_____ **27.** ___ pipe is steel pipe manufactured by drawing flat steel strips through a die to form a cylindrical shape and then electric butt-welding the seam to create a leakproof joint.
  A. Butt-welded
  B. Continuous weld
  C. Welded
  D. all of the above

_____ **28.** The unthreaded stem and handwheel of a ___ gate valve rise as the valve is opened to indicate the position of the wedge disk.
  A. nonrising stem-inside screw
  B. rising stem-inside screw
  C. rising stem-outside stem and yoke
  D. all of the above

_____ **29.** A ___ is a backflow prevention device that consists of a spring-loaded check valve that seals against an atmospheric outlet when water is turned on.
   A. backwater valve
   B. sillcock
   C. hose bibb
   D. vacuum breaker

_____ **30.** Steel pipe is available in nominal pipe sizes ranging from ___".
   A. ⅛ to 12
   B. ½ to 12
   C. 1 to 15
   D. 2½ to 18

_____ **31.** A ___ nipple is threaded its entire length.
   A. shoulder
   B. close
   C. wide
   D. no-shoulder

_____ **32.** A ___ fitting is a pipe fitting in which the dimension of at least one opening is smaller than other openings.
   A. reducing
   B. straight
   C. standard
   D. none of the above

_____ **33.** A ___ valve is designed to control fluid flow by partially opening or closing the valve.
   A. throttling
   B. shutoff
   C. full-way
   D. rated

_____ **34.** A(n) ___ valve is a non-rated globe valve with a side port in the valve body which is used to drain fluid from the outlet side of the valve.
   A. angle
   B. stop-and-waste
   C. backwater
   D. none of the above

_____ **35.** A ___ valve permits fluid flow in only one direction and closes automatically to prevent backflow.
   A. butterfly
   B. globe
   C. gate
   D. check

————————————— **36.** ___ valves have a split- or solid-wedge disk.
A. Ball
B. Gate
C. Globe
D. Backwater

————————————— **37.** A ___ relief valve is a safety device used to protect against the development of excessive temperature and/or pressure in a water heater.
A. pressure
B. T&P
C. temperature
D. none of the above

————————————— **38.** A ___ valve is an automatic device used to convert high and/or fluctuating inlet water pressure to a lower or constant outlet pressure.
A. backwater
B. butterfly
C. check
D. pressure-reducing

————————————— **39.** The threaded stem of a ___ gate valve rises as the valve is opened.
A. nonrising stem-inside screw
B. rising stem-inside screw
C. rising stem-outside stem and yoke
D. all of the above

————————————— **40.** A hinged disk or flapper within the body of a ___ backwater valve prevents backflow of sewage into a building.
A. lift check
B. swing check
C. ball
D. none of the above

————————————— **41.** A ___ valve is a full-way valve used to regulate fluid flow in which a threaded stem raises and lowers a wedge-shaped disk that fits against a valve seat within the valve body.
A. ball
B. globe
C. gate
D. backwater

————————————— **42.** A ___ disk gate valve is a gate valve in which a one-piece solid bronze wedge fits against the valve seat to restrict fluid flow.
A. single
B. split-wedge
C. solid-wedge
D. none of the above

_____ **43.** A(n) ___ is a globe valve in which the inlet and outlet are at 90° to each other.
    A. angle
    B. stop-and-waste
    C. backwater
    D. none of the above

_____ **44.** A ___ cock is installed on the water main to which the water service of a building is connected.
    A. curb
    B. corporation
    C. gas
    D. sill

_____ **45.** The disk in a ___ valve moves vertically within the valve body to prevent backflow.
    A. globe
    B. butterfly
    C. lift check
    D. swing check

_____ **46.** A ___ is a valve through which water or gas flow is controlled by a circular core or plug that fits closely in a machined seat.
    A. core cock
    B. gate valve
    C. globe valve
    D. none of the above

Completion

_____ **1.** A(n) ___ resin can be heated and reformed repeatedly with little or no degradation in physical characteristics.

_____ **2.** PVC DWV pipe and fittings are ___-colored.

3. The maximum developed length of rigid plastic drainage, waste, and vent piping is ___'.

_____ **4.** A(n) ___ is a device used to measure and indicate water flow.

_____ **5.** For any given diameter of copper tube, the outside diameter of all copper tube types is the same, which is ___" larger than the nominal or standard size.

_____ **6.** A(n) ___ is a device fastened to the ends of pipes to make connections between individual pipes.

_____ **7.** The ___ process is a PEX manufacturing process in which silane molecules are bonded to polyethylene molecules during the manufacturing process.

_____ **8.** A(n) ___ resin cannot be remelted after it is formed and cured in its final shape.

_____ 9. ___ copper tube is drawn copper tube that is heated to a specific temperature and cooled at a predetermined rate to impart desired strength and hardness characteristics.

_____ 10. ___ pipe is steel pipe that is cleaned and dipped into a hot molten zinc bath to create a protective coating.

_____ 11. A(n) ___ valve is a valve that meets or exceeds engineering criteria for the normal pressure range of the fluids contained within the system it is controlling.

_____ 12. A(n) ___ is a short piece of pipe, typically less than 12″ in length, with threads on each end.

_____ 13. A(n) ___ valve is used to control fluid flow by means of a pliable, circular disk that is compressed against a valve seat through which water flows.

_____ 14. A(n) ___ globe valve has a full-size valve seat opening.

_____ 15. ___ pipe is steel pipe that is coated with varnish to protect it against corrosion.

_____ 16. Valve bodies for 2½″ and larger valves are typically manufactured from cast ___ and have bronze internal components.

_____ 17. A(n) ___ is a valve with integral external threads installed on the exterior of a building for attachment of a garden hose.

_____ 18. Backflow is prevented through a(n) ___ check valve through the use of a hinged disk within the valve body.

_____ 19. A(n) ___ is a valve with hose threads that is installed on a tank to drain and/or flush the tank.

_____ 20. Water flow through a water service is measured using a(n) ___.

_____ 21. A(n) ___ is installed on a water service to turn on or off the potable water supply.

_____ 22. A(n) ___ valve consists of a rotating disk that seats against a resilient material within the valve body.

Short Answer

1. List five advantages that plastic pipe and fittings have over other piping materials.

_____

_____

_____

_____

_____

2. Explain why the overall cost of installing PVC pipe and fittings is greater than installing ABS pipe and fittings.

_____

_____

_____

_____

_____

_____

3. Describe the three PEX manufacturing processes.

_____

_____

_____

_____

_____

_____

_____

4. Describe the extrusion and piercing process for manufacturing copper tube.

_____

_____

_____

_____

_____

_____

_____

5. Describe the centrifugal casting process used to manufacture cast iron soil pipe.

_____

_____

_____

_____

_____

_____

6. List two advantages of grooved joint fittings over threaded pipe fittings.

_____

_____

_____

_____

_____

_____

7. Describe the operation of a disc water meter.

_____

_____

_____

_____

8. Describe the operation of a turbine water meter.

_____

_____

_____

_____

_____

9. Describe the operation of a compound water meter.

_____

_____

_____

_____

_____

Matching

ABS DWV Fittings

_____ **1.** Cap

_____ **2.** 90° elbow

_____ **3.** 45° wye

_____ **4.** 60° elbow

_____ **5.** Long-radius T-Y

_____ **6.** 45° elbow

_____ **7.** Double sanitary tee

_____ **8.** 22½° elbow

_____ **9.** Sanitary tee

_____ **10.** 90° street elbow

_____ **11.** Long-turn 90° elbow

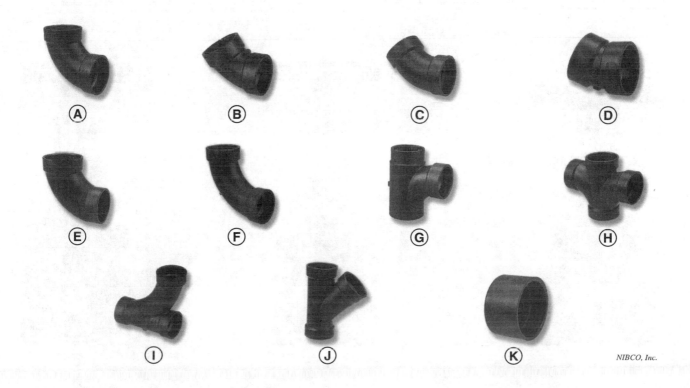

NIBCO, Inc.

No-Hub Cast Iron Soil Pipe Fittings

_____ **1.** Long-sweep ¼ bend

_____ **2.** Figure 1 fitting

_____ **3.** ¼ bend

_____ **4.** Sanitary tee

_____ **5.** Short-sweep ¼ bend

_____ **6.** Wye

_____ **7.** ¹⁄₁₆ bend

_____ **8.** ⅙ bend

_____ **9.** ⅛ bend

_____ **10.** Sanitary cross

_____ **11.** Figure 5 fitting

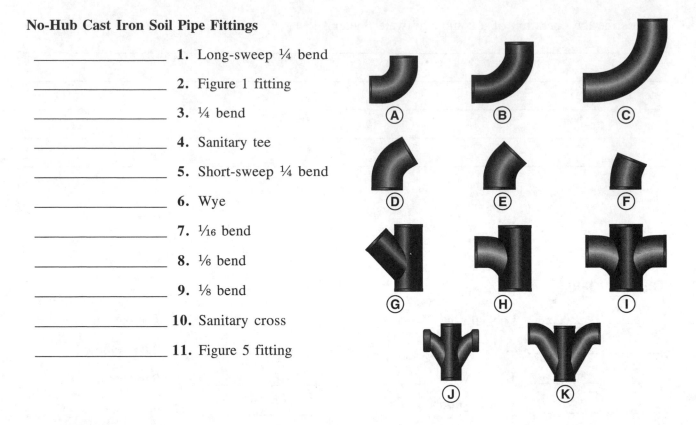

Gate Valves

_____ **1.** Nonrising stem-inside screw

_____ **2.** Rising stem-OS & Y

_____ **3.** Rising stem-inside screw

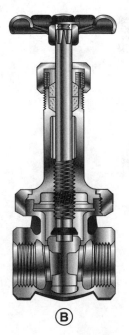

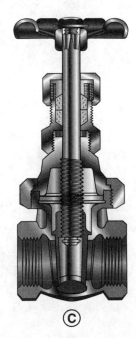

Globe Valves

_____ **1.** Gland

_____ **2.** Valve body

_____ **3.** Identification plate

_____ **4.** Disk

_____ **5.** Disk holder

_____ **6.** Annular ring seat

_____ **7.** Handwheel

_____ **8.** Stem

_____ **9.** Packing nut

_____ **10.** Union bonnet

_____ **11.** Packing

_____ **12.** Disk locknut

Ball Valves

_____ **1.** Ball

_____ **2.** Valve body

_____ **3.** Thrust washer

_____ **4.** Seat ring

_____ **5.** Packing gland

_____ **6.** Handle nut

_____ **7.** Body end piece

_____ **8.** Stem

_____ **9.** Packing

_____ **10.** Handle

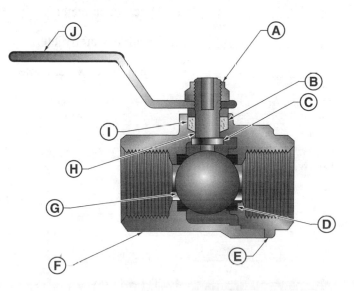

Pressure-Reducing Valves

_____ **1.** Data plate

_____ **2.** Integral strainer

_____ **3.** Valve body

_____ **4.** Inlet

_____ **5.** Outlet

_____ **6.** Adjustment screw

_____ **7.** Valve bonnet

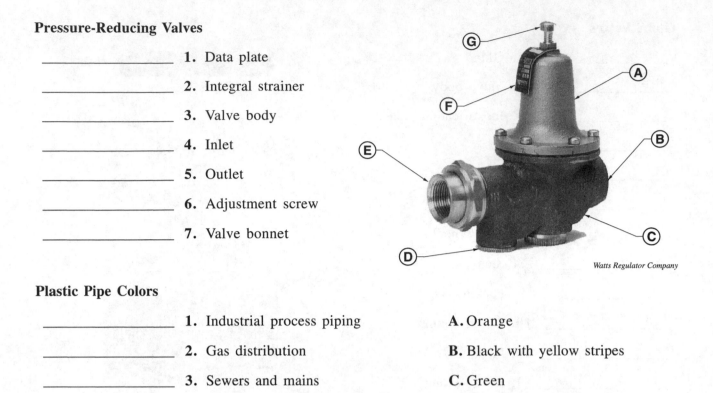

Watts Regulator Company

Plastic Pipe Colors

_____ **1.** Industrial process piping

_____ **2.** Gas distribution

_____ **3.** Sewers and mains

_____ **4.** Hot and cold water distribution

_____ **5.** Fire sprinklers

_____ **6.** Water distribution

_____ **7.** Drainage, waste, and vent piping

A. Orange

B. Black with yellow stripes

C. Green

D. Dark gray

E. Clear

F. Black

G. Light blue

Water Meters

_____ **1.** Disc water meter

_____ **2.** Turbine water meter

_____ **3.** Compound water meter

A. Used in buildings in which there is a large fluctuation of water flow

B. Used to measure water flow through small water services

C. Used to measure large and constant volumes of water in buildings

Drawn Copper Tube Colors

_____ **1.** Type K

_____ **2.** Type L

_____ **3.** Type M

_____ **4.** Type DWV

A. Red

B. Yellow

C. Green

D. Blue

Activities

Identifying Pipe Fittings

Pipe fittings are available in many standard configurations, including elbows, tees, wyes, and crosses. Most pipe fittings have at least two openings. Straight fittings are identified by their nominal size and configuration, such as ½″ 90° elbow or ¾″ tee. When identifying reducing fittings, the sizes of the inlet and outlet of the run are listed first, followed by the size of other outlets. Identify the following pipe fittings:

_____ **1.** Fitting A   _____ **4.** Fitting D

_____ **2.** Fitting B   _____ **5.** Fitting E

_____ **3.** Fitting C   _____ **6.** Fitting F

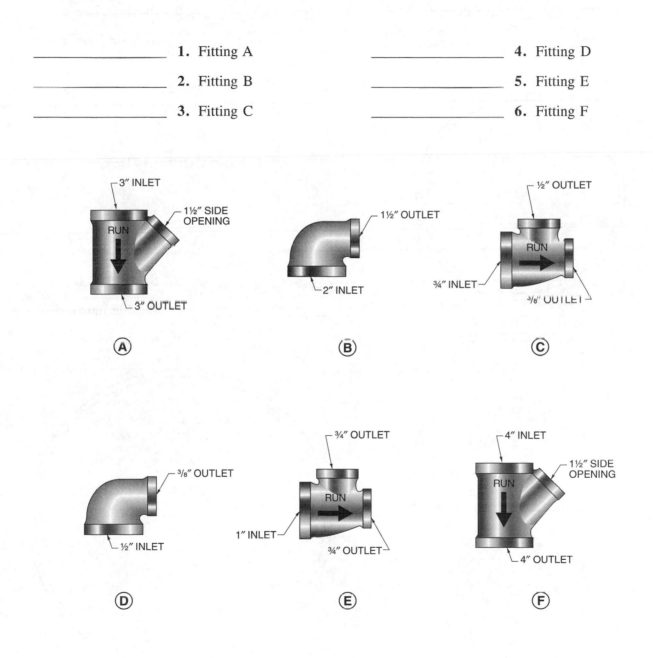

Determining Copper Tube Sizes

For any given diameter of copper tube, the outside diameter is the same for all copper tube types. The inside diameter of copper tube is determined by wall thickness. Measure the copper tube profiles to determine the nominal size, outside diameter, and inside diameter.

Example

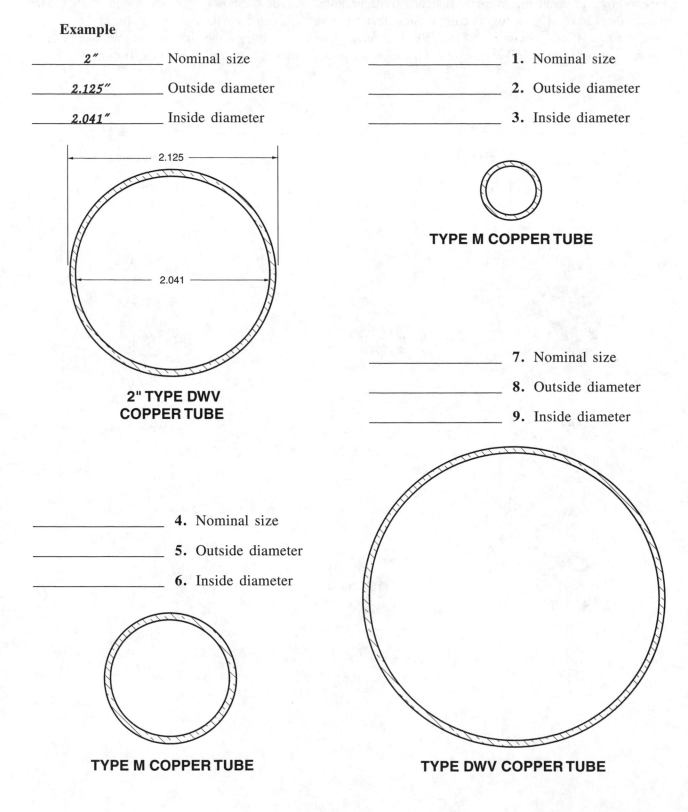

_2"_	Nominal size
_2.125"_	Outside diameter
_2.041"_	Inside diameter

2.125

2.041

**2" TYPE DWV
COPPER TUBE**

_____ **1.** Nominal size

_____ **2.** Outside diameter

_____ **3.** Inside diameter

TYPE M COPPER TUBE

_____ **7.** Nominal size

_____ **8.** Outside diameter

_____ **9.** Inside diameter

_____ **4.** Nominal size

_____ **5.** Outside diameter

_____ **6.** Inside diameter

TYPE M COPPER TUBE

TYPE DWV COPPER TUBE

_____ **10.** Nominal size

_____ **11.** Outside diameter

_____ **12.** Inside diameter

_____ **13.** Nominal size

_____ **14.** Outside diameter

_____ **15.** Inside diameter

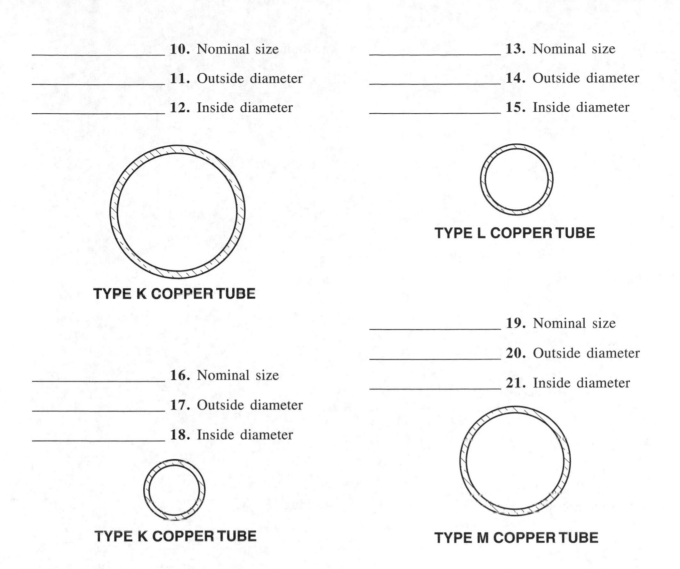

TYPE K COPPER TUBE

TYPE L COPPER TUBE

_____ **16.** Nominal size

_____ **17.** Outside diameter

_____ **18.** Inside diameter

_____ **19.** Nominal size

_____ **20.** Outside diameter

_____ **21.** Inside diameter

TYPE K COPPER TUBE

TYPE M COPPER TUBE

Sketching

Sketch the malleable iron threaded fittings in the space provided.

1. 90° Elbow

2. 45° Elbow

3. Tee

4. Reducer

5. Union

6. Cap

7. 90° Street Elbow

8. Plug

Projects

1. Browse the web sites of plumbing fitting and piping manufacturers to determine the types of products they offer. Locate their on-line product catalogs or request printed catalogs from the manufacturers. Review the catalogs, identifying the various sizes and configurations of fittings available from the manufacturers.

2. Obtain obsolete or damaged valves from a plumbing contractor or supply house. Cut the valves using a hacksaw or other appropriate saw to expose the internal components of the valves. **Warning:** Use appropriate personal protective equipment when cutting the valves. Identify the components and note the materials from which the components were manufactured.

Name_____ Date _____

True-False

   T   F   **1.** Steel tapes are available in 25′, 50′, and 100′ lengths.

   T   F   **2.** A torpedo level is a short level that uses an air bubble located inside a liquid-filled vial to establish plumb and level references.

   T   F   **3.** A laser torpedo level is equipped with an accurate laser beam.

   T   F   **4.** Carpenter's levels are available in 10′ lengths.

   T   F   **5.** A pipe laser level is a laser level that emits a rotating laser beam and is mounted inside a pipe or along the top edge of a pipe to establish level and grade references.

   T   F   **6.** A chalk line is pulled tightly between two points and the line is carefully pulled at a 45° angle to the surface and snapped, leaving a line of chalk on the surface.

   T   F   **7.** Pencils and soapstone are used to mark pipe and fittings.

   T   F   **8.** A clamp pipe vise holds pipe using a chain that is placed over the pipe.

   T   F   **9.** Handsaws are used to manually cut metal, wood, and plastic.

   T   F   **10.** The maximum-size hole for the desired pipe should be drilled in structural members to ensure their integrity and minimize loss of structural strength.

   T   F   **11.** Reinforced concrete is reinforced with steel reinforcement such as steel bars and welded wire fabric.

   T   F   **12.** Hammer-drills and rotary hammers rotate and drive simultaneously, causing the bit to spin and hammer at the same time.

   T   F   **13.** Two factors affect the drilling rate with a hammer-drill—speed and angle.

   T   F   **14.** Rotary hammers operate at a lower rpm than hammer-drills and should not be used to drill wood or metal.

   T   F   **15.** Rotary hammers require slotted drive system or slotted drive shank chucks and bits.

   T   F   **16.** A scissors cutter is a ratcheting plastic pipe cutter in which the handles are squeezed together to cut the pipe.

T   F   **17.** A tubing cutter is a cutting tool used to make a square cut on tube; it consists of a cutting frame and guide edges.

T   F   **18.** Soldered joints are commonly used on copper water pipe and in applications where the service temperature does not exceed 250°F.

T   F   **19.** Flared joint fittings are used with types L and M annealed copper tube and are typically used for underground water service applications.

T   F   **20.** A different size flaring tool is needed for each size of copper tube.

T   F   **21.** When bending copper tube, ensure that the tube does not shrink, reducing the flow capacity of the tube.

T   F   **22.** Steel pipe is cut and reamed with a pipe cutter before it is threaded or grooved.

T   F   **23.** A pipe cutter does not remove any metal; instead, the wheel squeezes the metal and forces it ahead of the cutter until the pipe is cut through.

T   F   **24.** The dies of a nonratcheting pipe threader are rotated completely around steel pipe to turn the die head, and they require adequate clearance for the handles.

T   F   **25.** A drophead pipe threader is a nonratcheting pipe threader in which the entire die head is replaced with another die head for different pipe sizes.

T   F   **26.** Smaller pipe sizes typically have more threads per inch than larger pipe sizes.

T   F   **27.** A universal die head is adjustable and is used to thread several pipe sizes with different dies.

T   F   **28.** A chain wrench is a pipe wrench with a serrated jaw, which uses a chain to secure the pipe in position.

T   F   **29.** Axe wrenches are available as straight wrenches or offset wrenches.

T   F   **30.** Basket strainer wrenches are a set of three wrenches used to tighten the nut that secures the kitchen sink basket strainer assembly to the sink.

T   F   **31.** A mechanical test plug is a test plug that is secured to pipe using mechanical means to prevent air or water leakage during an air or hydrostatic test.

T   F   **32.** An inflatable test plug is inserted into pipe and secured in position by inflating the plug with water.

T   F   **33.** When pressurizing a piping system, do not stand in front of an opening sealed with a cap or test plug.

T   F   **34.** Long-handle shovels provide greater leverage; short-handle shovels are convenient to use in areas with limited access.

Multiple Choice

_____ **1.** A ___ is a measuring tape consisting of a long continuous steel strip that is graduated in standard increments.
  A. coil
  B. reel
  C. blade lock
  D. steel tape

_____ **2.** A plumb bob is a cone-shaped metal weight fastened to a string that is used to establish a true ___.
  A. transfer plane
  B. horizontal plane
  C. plumb line
  D. base

_____ **3.** A ___ level is a level that uses a telescope that can be adjusted vertically and horizontally to establish straight-line references.
  A. pipe
  B. transit
  C. digital
  D. sensor

_____ **4.** The head of the laser transit level revolves at a maximum rate of ___ revolutions per minute.
  A. 160
  B. 260
  C. 360
  D. 460

_____ **5.** A ___ is an assembly for holding pipe in place during cutting, threading, or grooving operations.
  A. pipe vise
  B. grip vise
  C. chain grip
  D. slip-joint

_____ **6.** A ___ saw is a handsaw used to cut openings in relatively thin materials.
  A. hack
  B. band
  C. portable band
  D. compass

_____ **7.** Reciprocating saw blades with ___ teeth per inch are used for rough-in work.
  A. 2 to 4
  B. 4 to 6
  C. 6 to 8
  D. 8 to 10

_____ **8.** A ___ saw is a general-purpose electric or battery-powered saw used in wood-frame construction to cut holes for pipe over 2″ size.
  A. reciprocating
  B. portable band
  C. keyhole
  D. chain

_____ **9.** Hacksaw blades with ___ teeth per inch are typically used to cut steel pipe.
  A. 24
  B. 34
  C. 42
  D. 43

_____ **10.** Sheet metal ___ are a hand tool used to make straight cuts in light-gauge sheet metal.
  A. razors
  B. scissors
  C. snips
  D. clips

_____ **11.** Aviation ___ are a hand tool used to make straight or angled cuts in light-gauge sheet metal.
  A. snips
  B. clips
  C. scissors
  D. razors

_____ **12.** A ___ is a drill used to drill holes larger than ⅝″ diameter in concrete or masonry.
  A. rotary hammer
  B. trigger hammer
  C. push
  D. heavy-duty angle

_____ **13.** A(n) ___ saw is a circular saw that is attached to a small table and pivots at the rear.
  A. blade
  B. chop
  C. rotating
  D. axe

_____ **14.** A(n) ___ cutter is a plastic pipe cutter used to make accurate and square cuts from the inside of the pipe.
  A. burr
  B. square
  C. external
  D. internal

_____ **15.** A(n) ___ tool is a hand tool with a conical tip which is inserted into the end of PEX tubing and expanded as the handles are pushed together.
   A. spreader
   B. opener
   C. expander
   D. amplifier

_____ **16.** A ___ gauge is a flat piece of sheet metal with standard-size notches that is used to verify the proper diameter of crimping rings.
   A. go/no-go
   B. no/no-go
   C. stop/no-stop
   D. go/no-stop

_____ **17.** Copper tubing cutters are available in a variety of sizes for cutting copper tube from ___".
   A. ¼ to 4
   B. ¼ to 5
   C. ¼ to 6
   D. ¼ to 7

_____ **18.** A deburring tool or ___ is used to remove the burr and ridge to prepare copper tube for joining and assembly.
   A. steamer
   B. ridger
   C. beamer
   D. reamer

_____ **19.** Copper tube ___ are used to clean the inside mating surfaces of copper fittings.
   A. abrasive cloths
   B. soft cloths
   C. fitting brushes
   D. cleaning brushes

_____ **20.** ___ is a copper tube joining method in which a nonferrous filler metal with a melting temperature of 840°F or more is used.
   A. Brazing
   B. Welding
   C. Lining
   D. Soldering

_____ **21.** The filler metal, or solder, must be heated to its ___ point to ensure that the metal will flow properly between the tube and fittings.
   A. boiling
   B. melting
   C. baking
   D. torching

_____ **22.** A torch assembly that uses ___ gas, a mixture of acetylene and liquefied petroleum gases, is commonly used as a heat source when soldering or brazing.
  A. PPAM
  B. PAMP
  C. AMPP
  D. MAPP

_____ **23.** Rolled groove joint copper fittings range from ___″ size and are used for aboveground potable water supply applications.
  A. ¼ to 4
  B. ½ to 4
  C. 1½ to 8
  D. 2½ to 8

_____ **24.** A tube end notcher is a manually operated cutter that cuts a curved notch and produces ___ dimples simultaneously on the branch tube.
  A. one
  B. two
  C. three
  D. four

_____ **25.** A(n) ___ flaring tool is driven into the end of copper tube with a hammer to form a flare.
  A. impact
  B. contact
  C. flex
  D. opposing

_____ **26.** A ___ flaring tool is a flaring tool in which a flare is formed by forcing a conical flare cone into the end of copper tube.
  A. yoke-and-chain
  B. yoke-and-screw
  C. yoke-and-spring
  D. yoke-and-collar

_____ **27.** A spring bender is a copper tube bending tool for free-form bending of annealed copper tube from size ___″.
  A. ¼ to ⅓
  B. ¼ to ⅔
  C. ¼ to ¾
  D. ¼ to 1

_____ **28.** A ratchet bender is a copper tube bending tool for bending ___″ drawn copper tube.
  A. ½ to ⅓
  B. ½ to ⅔
  C. ½ to ¾
  D. ½ to 1

_____ **29.** A ratchet cutter is a pipe cutter used to cut ___" size cast iron soil pipe.

    A. ½ to 8

    B. 1½ to 8

    C. 1½ to 9

    D. 2 to 10

_____ **30.** A chop saw equipped with an abrasive blade is used to cut ___" no-hub cast iron soil pipe.

    A. 1½ to 4

    B. 1½ to 6

    C. 2 to 4

    D. 2½ to 4

_____ **31.** Steel pipe is used for ___.

    A. water distribution

    B. sanitary waste and vent

    C. storm water drainage

    D. all of the above

_____ **32.** A three-way pipe threader is a nonratcheting pipe threader that holds ___ die sizes simultaneously.

    A. one

    B. two

    C. three

    D. four

_____ **33.** Threading and grooving dies must be liberally lubricated when cutting threads or grooves to ___.

    A. speed metal removal

    B. produce a smooth finish

    C. prevent metal chips from jamming the dies

    D. all of the above

_____ **34.** A pipe groover is an adjustable tool used to cut a standard groove in ___" steel pipe.

    A. 1 to 8

    B. 2 to 8

    C. 2 to 9

    D. 2 to 10

_____ **35.** Small-diameter pipe machines are designed for ___" pipe and smaller and can be mounted on a tripod.

    A. 1

    B. 2

    C. 3

    D. 4

_____ **36.** Pipe larger than ___″ is reamed with a half-round file.
    A. 1
    B. 2
    C. 3
    D. 4

_____ **37.** ___ dies are mounted in a die head.
    A. Two
    B. Three
    C. Four
    D. Five

_____ **38.** A(n) ___ die head is a die head with a lever at the top that is raised to manually retract the dies and release the pipe when the thread has been cut to the desired length.
    A. quick-opening
    B. slow-opening
    C. automatic
    D. sensor

_____ **39.** A(n) ___ is a pipe machine accessory used to support long lengths of pipe and prevent the pipe from tipping the machine over.
    A. pipe clamp
    B. pipe support
    C. upright support
    D. all of the above

_____ **40.** A(n) ___ is a pipe machine accessory that transfers power from a pipe machine to a geared threader.
    A. universal drive shaft
    B. international drive shaft
    C. geared drive
    D. transfer shaft

_____ **41.** An adjustable wrench is a smooth-jawed general-purpose wrench used to tighten ___″ nuts.
    A. ¼ to 1
    B. ½ to 1
    C. ½ to 2
    D. ½ to 4

_____ **42.** A ___ wrench is a hand tool used to tighten chrome-plated pipe and fittings so that they will not be marred.
    A. cinch
    B. strip
    C. strap
    D. all of the above

_____ **43.** The size and type of pipe wrench used is determined by the ___.

    A. size of pipe being assembled

    B. approximate area in which work is performed

    C. personal preference for a certain type of wrench

    D. all of the above

Completion

_____ **1.** A(n) ___ is a measuring tool made of short, graduated sections fastened together to allow them to pivot and fold for convenient carrying or storage.

_____ **2.** A(n) ___ is a device used to establish accurate horizontal and vertical surfaces.

_____ **3.** A(n) ___ level uses a laser beam and receiver to establish level and plumb references.

_____ **4.** ___ lines are commonly used to lay out grid and wall lines on floors so that pipe can be accurately installed before walls are constructed.

_____ **5.** A(n) ___ pipe vise holds pipe using a yoke that is tightened down on the pipe with a threaded stem.

_____ **6.** A(n) ___ saw is an electric or battery-powered saw used to cut most metals, plastic, and irregularly shaped materials.

_____ **7.** Reciprocating saw blades with ___ teeth per inch or more are used for heavy gauge metal, tubing, and galvanized steel pipe.

_____ **8.** Hacksaw blades with ___ teeth per inch are used to cut plastic tubing.

_____ **9.** Portable ___ are used to cut circular openings in studs, joists, or wall coverings to accommodate pipe runs.

_____ **10.** A(n) ___ drill is an electric or battery-powered drill commonly used to drill holes squarely through studs or joists spaced 16″ or 24″ on-center.

_____ **11.** A(n) ___ drill is a piece of drilling equipment that consists of a drill head, mast, and base, and is used to cut holes up to 14″ diameter in concrete, concrete masonry units, and stone.

_____ **12.** A(n) ___ is a drill used to drill holes in concrete, steel, and wood using standard twist drills.

_____ **13.** A(n) ___ saw is a handsaw with teeth sharpened on both sides so that it can cut on both the forward and backward stroke to efficiently cut plastic pipe.

_____ **14.** A(n) ___ cutter is used to cut plastic pipe up to 1″ size.

_____ **15.** A(n) ___ tool is used to crimp, or tighten, a crimping ring securely around PEX tubing and fittings.

_____ **16.** Copper fittings larger than ___″ are cleaned with abrasive cloth.

_____ **17.** ___ is a copper tube joining method in which nonferrous filler metal with a melting temperature of less than 840°F is used.

_____ **18.** A(n) ___ is a tool used to form a small-radius groove on the end of copper tube.

_____ **19.** ___, or tee-drilling, is the process of extruding a branch collar, or tee, from the wall of copper tube.

_____ **20.** Spring, lever, and ratchet benders are used to bend ¼″–¾″ size ___ tube.

_____ **21.** The ___ bender is a copper tube bending tool for bending ¼″ and ⅜″ drawn or annealed copper tube.

_____ **22.** A T-handle torque wrench is recommended for tightening the nuts on no-hub cast iron soil pipe joints since the wrench applies ___ in-lb of torque and then releases.

_____ **23.** A pipe ___ is a conical-shaped tool that is rotated inside the end of steel pipe to remove burrs.

_____ **24.** A(n) ___ is a device used to apply oil to dies, and captures used oil in a reservoir for recirculation.

_____ **25.** A(n) ___ is a tool used to cut external threads in pipe.

_____ **26.** A die ___ is the part of a pipe threader that secures the dies in position and applies pressure to cut external threads in pipe.

_____ **27.** A(n) ___ is a portable power source fitted with a die head or universal drive shaft and is used to drive a geared threader or cut grooving tool.

_____ **28.** A(n) ___ is a piece of threading equipment that attaches to large-diameter pipe and is driven by a pipe machine or power drive.

_____ **29.** A(n) ___ is a hand tool with serrated, adjustable jaws that is used to assemble threaded pipe and fittings.

_____ **30.** A chain ___ is a chain wrench with two serrated jaws and a long handle, and is used to provide maximum leverage for turning large-diameter pipe.

_____ **31.** A(n) ___ wrench is a general-purpose pipe wrench used to assemble ⅛″–8″ pipe.

_____ **32.** A(n) ___ wrench is a hand tool used to tighten water supply connections to the faucet behind fixtures and slip nuts on fixture traps in hard-to-reach areas.

_____ **33.** Powdered ___ is sprinkled on the strap to improve the grip of the wrench on the smooth chrome surface.

_____ **34.** A(n) ___ wrench is a hand tool used to tighten the rim clamps that secure a kitchen sink or lavatory to the underside of a countertop.

_____ **35.** A(n) ___ is a reinforced mechanical plug that seals over the outside end of plastic, steel, copper, and cast iron soil pipe.

_____ **36.** A(n) ___ assembly is a job-built test device used to measure pressure within waste and vent, water, gas, air, and other piping systems.

_____ **37.** A(n) ___ is a U-shaped tube partially filled with water, and is used to measure pressure within a closed system.

_____ **38.** For some applications, such as overhead rainwater leaders, pipe must be lifted into position using a(n) ___ or hand chain hoist.

Matching

Drill Bits

_____ **1.** Multipiece bit

_____ **2.** Spade bit

_____ **3.** Solid bit

_____ **4.** Core bit

_____ **5.** Twist drills

_____ **6.** Self-feed bit

_____ **7.** Hole saw

A. Multipiece bit without the pilot bit with segmented teeth welded to the cutting edge, used for cutting concrete, concrete masonry units, and stone

B. Solid bit used to bore holes in wood studs and floors in 1″–4⅝″ diameter

C. General-purpose solid bits commonly used to drill small holes in wood or light-gauge metal

D. Multipiece drill bit with segmented teeth formed along the cutting edge

E. Drill bit that consists of several components, including a mandrel, pilot bit, and hollow cylindrical bit

F. Drill bit manufactured from a solid cylindrical piece of steel, and has cutting lips along the edges of the spiral shaft of the bit

G. Solid bit used to bore holes in wood and plastic in ¼″–1½″ diameters, and has its cutting lips at the outside edges

Yoke Pipe Vises

_____ **1.** Jaw

_____ **2.** Adjustment handle

_____ **3.** Yoke

_____ **4.** Stem

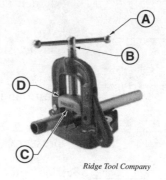

Ridge Tool Company

Chain Pipe Vises

_____ **1.** Adjustment handle

_____ **2.** Chain

_____ **3.** Tool tray

_____ **4.** Pipe vise stand

_____ **5.** Jaw

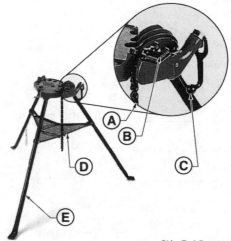

Ridge Tool Company

Plastic Pipe Saws and Cutters

_____ **1.** Universal

_____ **2.** Internal

_____ **3.** Chop

_____ **4.** Scissors

_____ **5.** Ratchet

Milwaukee Electric Tool Corp.
Ⓐ

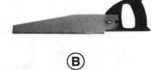
Ⓑ

Ridge Tool Company
Ⓒ

Ⓓ

Reed Manufacturing Co.
Ⓔ

Copper Tube Cutters

_____ **1.** ¼″ to 2″ size

_____ **2.** 1″ to 6″ size

_____ **3.** ¼″ to 1″ size

Ⓐ

Ⓑ

Ridge Tool Company

Ⓒ

Hand-Held Grinders

_____ **1.** Switch

_____ **2.** Wheel guard

_____ **3.** Grinding wheel

_____ **4.** Motor housing

_____ **5.** Side handle

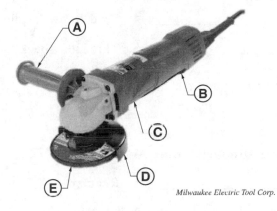

Milwaukee Electric Tool Corp.

Cast Iron Soil Pipe Cutting Tools

_____ **1.** Chop saw

_____ **2.** Ratchet cutter

_____ **3.** Hydraulic cutter

_____ **4.** Squeeze cutter

Ridge Tool Company

Ⓐ

Ridge Tool Company

Ⓑ

*Wheeler Mfg. Div. of
Rex International U.S.A. Inc.*

Ⓒ

Milwaukee Electric Tool Corp.

Ⓓ

Single-Wheel Steel Pipe Cutters

_____ **1.** Guide wheel

_____ **2.** Threaded shank

_____ **3.** Feed screw handle

_____ **4.** Frame

_____ **5.** Cutting wheel

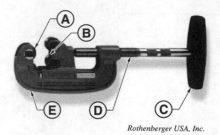

Rothenberger USA, Inc.

Multiple-Wheel Steel Pipe Cutters

_____ **1.** Frame

_____ **2.** Cutting wheel

_____ **3.** Hinge pin

_____ **4.** Feed screw handle

_____ **5.** Threaded shank

_____ **6.** Clamp

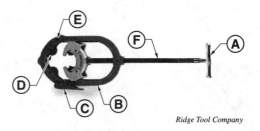

Ridge Tool Company

Large-Diameter Pipe Machines

_____ **1.** Reamer

_____ **2.** Foot switch

_____ **3.** Quick-opening die head

_____ **4.** Pipe cutter

_____ **5.** Accessory cabinet

_____ **6.** Floor stand

_____ **7.** Chuck

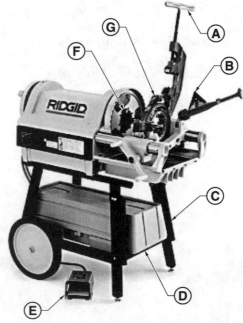

Ridge Tool Company

Pipe Wrenches

_____ **1.** Offset

_____ **2.** End

_____ **3.** Straight

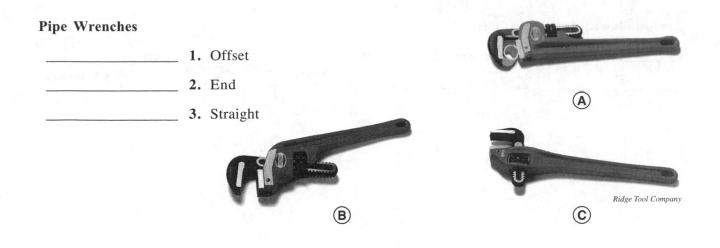

Ⓐ

Ⓑ

Ⓒ

Ridge Tool Company

Test Plugs and Caps

_____ **1.** Test cap

_____ **2.** Plastic test plug

_____ **3.** Long test plug

_____ **4.** Sealing test plug

_____ **5.** Short test plug

_____ **6.** Cast iron test plug

Ⓐ

Ⓑ

Ⓒ

Ⓓ

Ⓔ

Ⓕ

Smooth-Jawed Assembly Tools

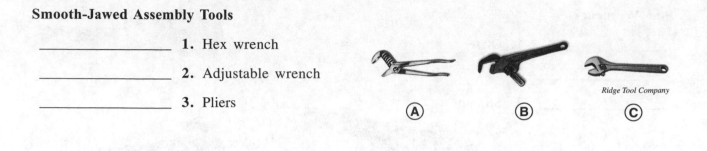

_____ **1.** Hex wrench

_____ **2.** Adjustable wrench

_____ **3.** Pliers

Ridge Tool Company

Ⓐ   Ⓑ   Ⓒ

Short Answer

1. Why is water used when drilling holes over 1″ diameter in reinforced concrete using a core bit?

_____

_____

_____

_____

2. Describe how copper tube and fittings are cleaned prior to applying flux.

_____

_____

_____

_____

_____

3. Why are the teeth of a universal saw sharpened on both sides?

_____

_____

_____

_____

4. What advantages are derived from the use of copper tube in a plumbing system?

_____

_____

_____

_____

5. Explain how a pipe laser level is used to establish a proper grade for a building sewer.

_____

_____

_____

_____

6. Explain how a tubing cutter is used to cut copper tube.

_____

_____

_____

_____

7. What is the purpose of a tube end notcher?

_____

_____

8. Describe the operation of a soil pipe assembly tool.

_____

_____

_____

_____

9. Why are pipe supports needed when working on long lengths of pipe with a small-diameter pipe machine?

_____

_____

_____

_____

10. Explain how a depth guide is used when drilling anchor holes to a predetermined depth.

_____

_____

_____

_____

11. List three reasons that threading dies must be liberally lubricated when cutting threads.

_____

_____

_____

_____

Activities

Effective Planning

Planning is essential to the efficient installation of piping materials, regardless of the type of pipe and fittings being installed. One element of good planning is ensuring that the proper tools and equipment are available for performing an operation.

1. Given a 20′ length of 4″ ABS DWV pipe, list the tools and equipment required to measure, mark, cut, and prepare three 5′ sections for use.

_____

_____

_____

2. Given a 5′ length of 4″ double-hub cast iron soil pipe, list the tools and equipment required to measure, mark, cut, and prepare two 2′-6″ sections for use.

_____

_____

_____

_____

_____

_____

3. Given a 21′ length of steel pipe, list the tools and equipment required to manually cut and thread the pipe.

_____

_____

_____

_____

_____

_____

Weight of Water in Pipes

Regardless of the pipe material, the weight of water in a pipe should be known before removing hangers or other supports when repairing plumbing systems. The weight of water is calculated using the formula:

Weight $= .34d^2 \times l$

where

Weight = weight of water (in lb)
.34 = constant
d = diameter of pipe (in in.)
l = length of pipe (in ft)

Determine the weight of water contained in the following sections of pipe.

_____ **1.** 4″ diameter × 10′ long

_____ **2.** 2″ diameter × 5′ long

_____ **3.** 8″ diameter × 12′ long

_____ **4.** 1¼″ diameter × 6′ long

_____ **5.** 2½″ diameter × 18′ long

Projects

1. Tools and equipment are typically available in several models and levels of quality to appeal to consumers and professionals. In general, professional-grade tools and equipment have greater durability and reliability than consumer-grade tools. Obtain tool and equipment manufacturer catalogs or visit their Web sites and compare and contrast specifications of common tools and equipment.

2. Several companies specialize in tools and equipment used in the plumbing trade. Contact tool and equipment manufacturers to request literature regarding their products. Review the product literature and compare and contrast the features of the tools and equipment. Develop a list of tools and equipment that an apprentice plumber might purchase to perform basic plumbing operations.

Name_____ Date _____

True -False

T   F   **1.** Before applying primer or solvent cement to plastic pipe and fittings, check for the proper interference fit.

T   F   **2.** Solder with a lead ingredient may be used for water supply applications.

T   F   **3.** Solvent cementing, or solvent welding, is the process in which plastic pipe and fittings are fused together by softening the adjoining surfaces with a chemical agent.

T   F   **4.** When brazing 1½″ and smaller copper tube, move the torch around the entire joint to bring it up to brazing temperature at the same time.

T   F   **5.** In the tee-pulling process, a branch collar, or tee, is extruded from the wall of copper tube.

T   F   **6.** PEX tubing manufactured using the silane or radiation process has the shape and thermal memory of Engel process PEX tubing.

T   F   **7.** Universal solvents may be used to join all types of plastic pipe and fittings.

T   F   **8.** Brazing filler metal is available in several silver, phosphorus, and copper alloys.

T   F   **9.** When soldering, apply heat to the entire circumference of a copper tube, with the flame perpendicular to the tube.

T   F   **10.** Bell-and-spigot cast iron soil pipe joints are mechanical joints, made with molded neoprene compression gaskets.

T   F   **11.** Before drilling an initial pilot hole during the tee-pulling process, collaring head forming pins should be in the open position.

T   F   **12.** When brazing, do not direct the flame at the filler metal as it is fed into the joint.

T   F   **13.** The American standard taper pipe thread (NPT) is tapered ¾″ per foot of thread length so that the pipe and fittings will thread together tightly to form a leakproof joint.

T   F   **14.** A trench bottom for installing underground drainage or waste pipe should be stable, relatively smooth, and properly graded.

T   F   **15.** An extension split pipe clamp is a plastic pipe hanger with an integral lug which assists in retaining the pipe in the hanger.

T   F   **16.** A C-shaped resilient rubber gasket conforms to the inside of pipe housing to seal pipe as housing segments are tightened.

T   F   **17.** Aboveground pipe must be properly hung and supported to keep pipe properly aligned and to prevent it from sagging and leaking.

T   F   **18.** Cast iron soil pipe, ABS, and PVC are commonly used for underground drainage and waste pipe.

T   F   **19.** A J-hanger is ¾″ strap iron with punched holes along its length, and is used to support horizontal pipe runs between joists.

T   F   **20.** A J-hook is a plastic pipe hanger that encompasses the entire circumference, and is driven in at an angle to trap the pipe between the wood structural member and the J-hook.

T   F   **21.** A bend support is a metal or reinforced plastic device that encloses a section of PEX tubing and provides rigid 90° bends.

T   F   **22.** Vertical pipe must be properly supported at close intervals to maintain alignment of the pipe.

T   F   **23.** Perforated strap iron is a hinged pipe clamp secured to a structural component with a threaded hanger rod.

Multiple Choice

_____ **1.** Brazing fluxes are ___-based, prevent formation of oxides, and remove oxides or other undesirable substances from copper tube.
    A. water
    B. oil
    C. petroleum
    D. nitrogen

_____ **2.** Procedures for safe handling of solvent cements, primers, and cleaners used in joining thermoplastic pipe and fittings include ___.
    A. avoiding breathing of solvent cement vapors
    B. proper disposal of all cloths used to clean excess solvent cement
    C. keeping chemicals away from sources of ignition
    D. all of the above

_____ **3.** The burr on the inside of galvanized steel pipe is removed using a ___ or a half-round file.
    A. pipe reamer
    B. cutting torch
    C. pipe cutter
    D. burr remover

       4. Compression joint fittings are used to make ___.

          A. aboveground connections that may need to be disassembled

          B. below-ground connections that should not be disassembled

          C. connections with cast iron soil waste pipe

          D. connections with PEX tubing

       5. ___ is (are) not a factor required to achieve a watertight and airtight solvent-cemented joint.

          A. The appropriate primer and solvent for the type of plastic pipe being joined

          B. A good interference fit

          C. Proper application of heat and solder

          D. Proper preparation and installation techniques

       6. Bell-and-spigot cast iron soil pipe compression gaskets are made from ___.

          A. cast iron

          B. neoprene rubber

          C. polystyrene

          D. silicone caulk

       7. DWV fittings are used for drain, waste, and venting applications with ___ tube.

          A. type K copper

          B. types L, M, and DWV copper

          C. PEX

          D. none of the above

       8. PEX tubing is used for ___ piping.

          A. water service

          B. hot water distribution

          C. cold water distribution

          D. all of the above

       9. ___ pipe is available in no-hub and bell-and-spigot patterns.

          A. Copper

          B. PVC

          C. Cast iron soil

          D. Steel

      10. ___ is (are) not a component of a cut groove pipe joint.

          A. A groove cut near the end of the galvanized steel pipe

          B. A housing and gasket

          C. Thread sealant

          D. Nuts and bolts

      11. Improperly installed pipe could result in ___.

          A. sagging joints

          B. joint leakage or breakage

          C. liquid or solid waste trapped within the pipe

          D. all of the above

_____ **12.** A ___ is a chemical agent that penetrates and softens the surface of plastic pipe and fittings.
      A. solvent cement
      B. primer
      C. cleaner
      D. reamer

_____ **13.** When making a rolled groove joint, the groove is formed by pressing a steel ___ into the copper tube wall and deforming the tube.
      A. blade
      B. drill bit
      C. die
      D. tube cutter

_____ **14.** A ___ is a pipe hanger that grasps the pipe but allows for thermal expansion of the pipe.
      A. perforated strap
      B. J-hanger
      C. talon
      D. wrap-around clamp

_____ **15.** In wood-frame construction, ___ are typically used to secure pipe anchors and supports to structural components.
      A. plastic inserts
      B. expansion plugs
      C. nuts and bolts
      D. nails and screws

_____ **16.** For underground stack bases, a ___ must be placed under the fitting at the base of the stack.
      A. pipe hanger
      B. brick or solid concrete block
      C. bend support
      D. riser clamp

_____ **17.** Flared joint fittings are used with type ___ annealed copper tube, and are typically used for underground water service applications.
      A. K
      B. L
      C. M
      D. all of the above

_____ **18.** Malleable iron threaded fittings are tapped with ___ pipe threads when the fittings are manufactured.
      A. female
      B. external
      C. tapered
      D. male

_____ **19.** In steel-frame construction, clamps are used to fasten pipe ___ or supports to the framing members.
 A. fittings
 B. hangers
 C. links
 D. joints

Completion

_____ **1.** Solvent cements should be applied when the surface temperature is above ___°F.

_____ **2.** The basic components of a copper rolled groove joint are the rolled groove near the end of the copper tube and a coupling composed of a two-piece housing, ___, and bolts and nuts to secure the coupling in position.

_____ **3.** A(n) ___ is a measuring device commonly included with a roll grooving tool, which is wrapped around the circumference of the groove to accurately measure groove diameter.

_____ **4.** ABS, ___, and CPVC plastic pipe and fittings are joined by solvent cementing.

_____ **5.** ___ is a condition that occurs when two metals are rubbed together under a load, resulting in erosion of the metal caused by friction.

_____ **6.** ___, or pipe dope, is material applied to male pipe threads to ensure an airtight joint.

_____ **7.** A(n) ___, such as a U-hook or J-hook, is a one-piece plastic or steel pipe hanger secured to horizontal structural components using nails or screws.

_____ **8.** Plumbing codes require that drainage, waste, and vent piping be ___-tested to ensure that the piping is leakproof.

_____ **9.** Pipe hangers and supports are anchored to structural components of a building, such as beams, studs, or joists, to ensure proper stability, ___, and alignment of pipe.

_____ **10.** PEX tubing manufactured using the Engel process has shape or ___, meaning that the tubing returns to its original shape after being expanded.

_____ **11.** Galvanized steel pipe is commonly used for large-diameter water supply and water distribution piping, and is joined using ___ pipe joints.

_____ **12.** Steel pipe and fittings are used for water distribution, sanitary waste and vent, ___ drainage, and gas piping systems.

_____ **13.** A ___ bend support is a metal or reinforced plastic device that encloses a section of PEX tubing and provides a nailing plate for stub-out applications.

_____ **14.** A(n) ___ is a pipe hanger used to support horizontal pipe and consists of a suspended U-shaped bracket with holes in the upper ends to receive a pin and an inverted bracket.

_____ **15.** A(n) ___ is a chemical agent that cleans and softens the surface of plastic pipe and fittings and allows solvent cement to penetrate more effectively into the pipe.

_____ **16.** A(n) ___ clamp is a plastic pipe hanger that fully encompasses the circumference of the plastic pipe.

_____ **17.** A(n) ___ is a one-piece pipe clamp secured to a stud with nails or screws.

_____ **18.** ___ is the temperature range within which solder is neither completely solid nor completely liquid, and is the working temperature range for a particular type of solder.

_____ **19.** A(n) ___ fit is a fit in which the diameter of the internal member, such as plastic pipe, is slightly larger than the smallest diameter of the external member, such as a fitting.

_____ **20.** Brazed joints in copper tube are made by heating the joint at temperatures greater than ___°F and adding brazing filler metal.

_____ **21.** A(n) ___ is a pipe hanger used to support multiple horizontal pipes and consists of a channel bracket on which the pipes rest and straps that secure the pipes in position.

_____ **22.** A(n) ___ is a two-piece pipe clamp used to support vertical runs of steel and cast iron pipe and copper tube.

Short Answer

1. List five precautions that should be taken when working with solvent cements, primers, and cleaners.

_____

_____

_____

2. Briefly describe the proper preparation and installation of PVC pipe and fittings.

_____

_____

_____

_____

_____

_____

_____

_____

3. List the procedure for properly installing expanded PEX tubing.

_____

_____

_____

_____

_____

_____

4. List the procedure for properly installing crimped PEX tubing.

_____

_____

_____

_____

_____

_____

_____

_____

5. Explain why 95-5 or lead-free solder must be used on water supply piping.

_____

_____

_____

_____

6. List three advantages of lead-free solder over 95-5 solder.

_____

_____

_____

_____

7. Briefly describe the proper preparation and installation of copper tube and fittings.

_____

_____

_____

_____

_____

_____

_____

_____

_____

_____

_____

_____

_____

8. Explain the difference between BCuP and BAg classes of filler metals.

_____

_____

_____

_____

_____

_____

_____

_____

9. Why should the hex nuts on no-hub cast iron soil pipe clamp assemblies be alternately tightened when assembling a joint?

_____

_____

_____

_____

_____

10. Briefly describe the proper procedure for threading and assembling steel pipe joints.

_____

_____

_____

_____

_____

_____

_____

_____

_____

11. Describe the proper procedure for backfilling around underground pipe.

_____

_____

_____

_____

_____

_____

_____

_____

Matching

Aboveground Stack Base Support

_____ **1.** Cleanout

_____ **2.** Threaded rod

_____ **3.** Clevis hanger

_____ **4.** Riser clamp above floor

_____ **5.** No-hub coupling

_____ **6.** Trapeze hanger

Supporting Horizontal Pipe

_____ **1.** Perforated strap iron

_____ **2.** Wrap-around clamp

_____ **3.** Loop

_____ **4.** Clevis hanger

_____ **5.** Pipe hooks

_____ **6.** Double J-hanger

_____ **7.** Talons

_____ **8.** J-hook

_____ **9.** Single J-hanger

_____ **10.** Trapeze hanger

Sioux Chief Manufacturing Company, Inc.

Anchors and Accessories

_____ 1. Wall/ceiling flange

_____ 2. Wedge anchor

_____ 3. Plastic insert

_____ 4. Expansion anchor

_____ 5. Steel anchor with expansion plug

_____ 6. Toggle anchor

_____ 7. Upper attachment

_____ 8. Sleeve anchor

_____ 9. Beam C-clamp

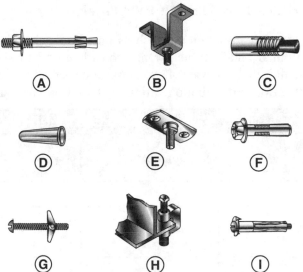

Soldering and Brazing Materials

_____ 1. 95-5 solder

_____ 2. Self-tinning flux

_____ 3. Lead-free solder

_____ 4. Flux

_____ 5. BAg class filler metal

_____ 6. BCuP class filler metal

A. Used to prevent formation of oxides or to facilitate removal of oxides from copper tube

B. Contains a tinning agent, and cleans, fluxes, and tins the surfaces of copper tube and fittings

C. Composed of 95% tin and 5% antimony

D. Composed of tin, copper, silver, bismuth, and selenium alloys

E. Contains copper and phosphorus; used for general piping applications

F. Contains 5%–15% silver; used for joining dissimilar metals

Copper Tube Joints

_____ 1. Soldered joints

_____ 2. Brazed joints

_____ 3. Flared joints

_____ 4. Rolled groove joints

A. Used on 2″–6″ copper water mains

B. Typically used on water supply piping

C. Used when soldering is not practical

D. Commonly used on medical gas or natural gas piping

Activities

Friction Loss Through PVC Fittings

For small installations, such as one-family dwellings, friction loss for water flow through PVC fittings is minimal and is typically not considered when sizing water distribution piping. For larger installations, friction loss for water flow must be considered when sizing the piping. Friction loss for water flow through PVC water distribution fittings is based on equivalent lengths of pipe as shown in the following table.

EQUIVALENT LENGTH OF PIPE PVC PIPE FITTINGS				
Pipe Size*	90° Elbow†	45° Elbow†	Through Tee Run†	Through Tee Branch†
½	1.5	8	1	4
¾	2	1	1.4	5
1	2.5	1.4	1.7	6
1¼	4	1.8	2.3	7
1½	4	2	2.7	8
2	6	2.5	4.3	12
2½	8	3	5.1	15
3	8	4	6.3	16
4	12	5	8.3	22
6	18	8	12.5	32
8	22	10	16.5	38
10	26	13.5	17.5	57
12	32	15.5	20	67

* in in.
† in ft

For example, when a 2½″ 90° elbow and 2½″ 45° elbow are installed in a run of water distribution piping, the equivalent length of pipe is 11′ (8′ + 3′ = 11′). Determine the equivalent length of pipe for the following installations:

_____ **1.** Equivalent length of pipe = ___′.

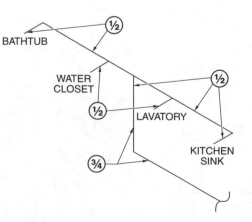

_____ **2.** Equivalent length of pipe = ___′.

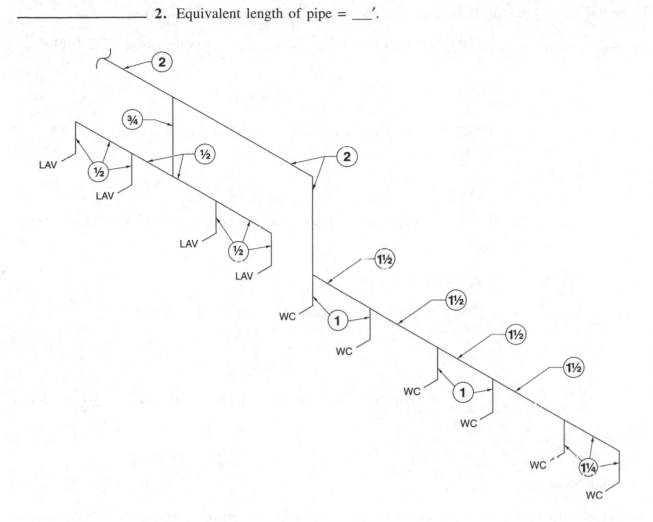

Copper Tube Expansion and Contraction

Copper tube, like other piping material, is subject to expansion and contraction due to temperature changes. The average coefficient of expansion of copper is .0000094 in./in./°F between 70°F and 212°F. The formula for calculating expansion and contraction of copper tube is:

Expansion (or contraction) = °F × L × 12 in./ft × .0000094

where

Expansion (or contraction) = expansion or contraction of copper tube (in in.)
°F = temperature change (in °F)
L = length (in ft)
12 in./ft = constant
.0000094 = constant

For example, 100′ of copper tube will expand 1.128″ when the tube is heated from 85°F to 185°F.
Expansion = °F × L × 12 in./ft × .0000094
Expansion = 100°F × 100′ × 12 in./ft × .0000094
Expansion = **1.128″**

Determine the expansion or contraction of the copper tube in the following examples.

_____ **1.** 100′ of copper tube will expand ___″ when heated from 70°F to 140°F.

_____ **2.** 50′ of copper tube will expand ___″ when heated from 100°F to 200°F.

_____ **3.** 100′ of copper tube will contract ___″ when cooled from 175°F to 100°F.

_____ **4.** 75′ of copper tube will expand ___″ when heated from 115°F to 175°F.

_____ **5.** 100′ of copper tube will contract ___″ when cooled from 212°F to 70°F.

Steel Pipe Makeup

The fitting allowance must be considered when determining the proper length of pipe to be cut. The fitting allowance is the measurement from the end of a piece of pipe when it is properly installed to the center of the fitting. The fitting allowance is subtracted from the overall length of an assembly to determine the proper length of pipe.

1. Using the drawing, list the parts required, including name and size.

_____

_____

_____

2. Using the drawing, list the tools required.

_____

_____

_____

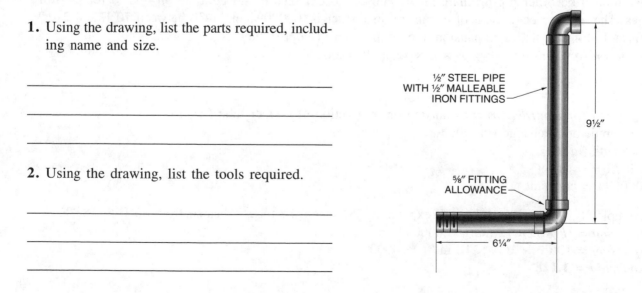

½″ STEEL PIPE WITH ½″ MALLEABLE IRON FITTINGS

9½″

⅝″ FITTING ALLOWANCE

6¼″

Pipe Threads

The American standard taper pipe thread is a standard pipe thread used for connecting water, gas, and steam pipes in which the adjoining sides of the pipe threads are at 60° to each other. The pipe threads are tapered ¾″ per foot of thread length to ensure a leakproof joint. Characteristics such as the number of threads per inch, thread length, number of threads, and total thread makeup vary with the nominal pipe size.

AMERICAN STANDARD TAPER PIPE THREAD CHARACTERISTICS				
Nominal Pipe Size*	Threads Per Inch	Approximate Thread Length	Approximate Number of Threads	Approximate Total Thread Makeup*
⅛	27	⅜	10	¼
¼	18	⅝	11	⅜
⅜	18	⅝	11	⅜
½	14	¾	10	⁷⁄₁₆
¾	14	¾	10	½
1	11½	⅞	10	⁹⁄₁₆
1¼	11½	1	11	⁹⁄₁₆
1½	11½	1	11	⁹⁄₁₆
2	11½	1	11	⅝
2½	8	1½	12	⅞
3	8	1½	12	1
3½	8	1⅝	13	1¹⁄₁₆
4	8	1⅝	13	1¹⁄₁₆
5	8	1¾	14	1³⁄₁₆
6	8	1¾	14	1⁹⁄₁₆
8	8	1⅞	15	1⁵⁄₁₆
10	8	2	16	1½
12	8	2⅛	17	1⅝

* in in.

For each of the following thread profiles, determine the number of threads per inch and the approximate length of threads to be cut.

_____ **1.** Threads per inch for ⅜″ nominal size pipe – ___

_____ **2.** Approximate thread length for ⅜″ nominal size pipe = ___″

_____ **3.** Threads per inch for ¾″ nominal size pipe = ___

_____ **4.** Approximate thread length for ¾″ nominal size pipe = ___″

_____ **5.** Threads per inch for 1½″ nominal size pipe = ___

_____ **6.** Approximate thread length for 1½″ nominal size pipe = ___″

_____ **7.** Threads per inch for 3″ nominal size pipe = ___

_____ **8.** Approximate thread length for 3″ nominal size pipe = ___″

Projects

1. Trade associations, such as the Plastic Pipe and Fittings Association (PPFA) and Copper Development Association (CDA), provide information about the products they represent including specification sheets and installation procedures. Visit trade association Web sites or contact the associations to obtain information.

2. Anchors and accessories are typically available from home improvement stores and outlets. Visit a local home improvement store or outlet and obtain anchors and accessories that may be used in plumbing installations. Identify applications for which each of the anchors or accessories may be used.

Name_____ Date _____

True-False

T	F	**1.**	A vertical pipe is any pipe or fitting that makes an angle of 45° or less with the vertical plane.

T   F   **2.** Sanitary drainage piping is sized according to the drainage fixture unit system.

T   F   **3.** Sanitary drainage piping conveys wastewater and waterborne waste from the plumbing fixtures and appliances to the sanitary sewer.

T   F   **4.** No more than three water closets may drain into a 3″ horizontal drainage pipe.

T   F   **5.** Properly sized horizontal drainage pipe should be approximately three-fourths full of sewage to ensure proper scouring action.

T   F   **6.** Vertical pipes or stacks that empty into the horizontal drainage pipes are sized after the horizontal building sewer and building drain pipe are sized.

T   F   **7.** No soil or waste stack is permitted to be larger than the smallest horizontal branch connected to it.

T   F   **8.** An offset is used for soil or waste stacks if the stacks cannot continue vertically over their entire length.

T   F   **9.** A 1½″ pipe is required as a common vent for four lavatory traps.

T   F   **10.** Cleanouts provide access to an individual vent if a stoppage occurs.

T   F   **11.** Since sewer gas is foul-smelling, a vent terminal should be located at least 10′ from windows, doors, or other ventilation openings.

T   F   **12.** Waste pipes are typically sized larger than soil pipes in a building.

T   F   **13.** Most plumbing codes require all underground drainage pipes to be at least 2″ in diameter.

T   F   **14.** Vent pipes are sized based on the drainage fixture units connected to the vent pipe and the developed length of vent pipe.

T   F   **15.** Plus pressure is a pressure less than one atmosphere.

T   F   **16.** A stack that is offset 45° or more from the vertical plane is sized as though it is a straight vertical stack.

T   F   **17.** A partial vacuum may develop in the drainage system, affecting its discharge capacity.

T   F   **18.** Retarded flow in a drainage system is the result of improper atmospheric conditions, insufficient venting, or improper fitting installation.

T   F   **19.** Common venting is used on four similar fixtures with the same vertical drain heights.

T   F   **20.** A branch vent is a vent pipe connecting two or more individual vents with a stack vent or vent stack.

T   F   **21.** The storm water drainage system conveys rainwater and other precipitation to the storm sewer or other place of disposal.

T   F   **22.** When venting wall-hung or cabinet-set fixtures, such as sinks and lavatories, the trap discharges into the side opening of a sanitary drainage tee.

T   F   **23.** Horizontal drainage piping is graded ¾″ per foot length of pipe.

T   F   **24.** Cleanouts for underground piping must be closely watched during backfilling operations so the cleanouts remain intact and are not buried or damaged.

T   F   **25.** Any building in which plumbing is installed must have at least one 3″ or larger diameter stack vent or vent stack extended full size through the roof.

T   F   **26.** When drains are installed, flexing is installed around the drain body and sealed to the roof to form a watertight connection.

T   F   **27.** Rainwater leaders usually run along a cornice or are placed in vertical shafts constructed specifically for pipe.

T   F   **28.** Projected roof area is the primary factor in determining size of building storm drains and rainwater leaders.

T   F   **29.** The International Association of Plumbing and Mechanical Officials developed the Uniform Plumbing Code.

T   F   **30.** In addition to being foul-smelling, sewer gases contain chemical elements that, when combined with moist air, create acids that corrode vent pipes.

T   F   **31.** Roof terminals for vents must be sealed to the roof surface to prevent rainwater, snow, and other moisture around the pipe from leaking into the building.

T   F   **32.** Roof jackets or roof flanges are installed around vent terminals to make the roof watertight.

T   F   **33.** The distance between a fixture trap and vent is based on the fixture drain size.

T   F   **34.** When sizing offset stacks, the portion above the offset is sized first, followed by the offset, and then the portion below the offset.

T   F   **35.** Flat venting of water closets should be avoided since stoppage of the fixture drain can result in waste backing into the vent and plugging it.

Multiple Choice

_____ 1. Wastewater discharge rates of plumbing fixtures are based on 1 dfu being equal to ___ gal. of water per minute of waste discharge.
   A. 2½
   B. 3½
   C. 4½
   D. 7½

_____ 2. A ___ is a measure of the probable discharge of wastewater and waterborne waste into the drainage system by various plumbing fixtures.
   A. drainage fixture unit
   B. change in direction
   C. trap seal loss
   D. branch interval

_____ 3. Discharge capacities for different sizes of horizontal drainage pipes are typically found in ___.
   A. plumbing codes
   B. water tables
   C. fixture trap-to-vent distance tables
   D. common vent tables

_____ 4. A building drain should not be less than ___″ in diameter if it receives the discharge of three or more water closets.
   A. 2
   B. 3
   C. 4
   D. 8

_____ 5. In general, one branch interval equals ___ floor(s) of plumbing fixture drains.
   A. 1
   B. 2
   C. 3
   D. 4

_____ 6. A ___ vent is the principal artery of the vent system to which vent branches may be connected.
   A. wet
   B. back
   C. main
   D. relief

_____ 7. A yoke vent is a type of ___ vent that provides additional air circulation between drainage and vent systems.
   A. fixture unit
   B. back
   C. common
   D. relief

_____ **8.** A ___″ cleanout is typically the largest size cleanout installed in a sanitary drainage system.
     A. 3
     B. 4
     C. 5
     D. 6

_____ **9.** Drainage piping that is less than 3″ and changes in direction from vertical to horizontal requires a ___ fitting.
     A. long-turn T-Y
     B. short sweep ¼ bend
     C. long-turn 90° elbow
     D. 45° wye

_____ **10.** A properly installed vent system provides air circulation within sanitary drainage piping and ensures that trap seals will not be subject to a pressure of more than ___″ of water column.
     A. 1
     B. 2
     C. 3
     D. 4

_____ **11.** The most common individual vent is a ___ vent.
     A. roof
     B. continuous
     C. relief
     D. stack

_____ **12.** ___ pipes convey discharge containing fecal matter from water closets or similar fixtures.
     A. Vent
     B. Waste
     C. Storm
     D. Soil

_____ **13.** Changes in ___ affect the discharge capacities of drainage pipes and are considered in sizing tables.
     A. direction
     B. material
     C. grade
     D. all of the above

_____ **14.** No water closet is permitted to discharge its waste into a horizontal drainage pipe less than ___″ in diameter.
     A. 2
     B. 3
     C. 4
     D. 4½

_____ **15.** Building sewers must be at least ___″ in diameter.
  A. 1
  B. 2
  C. 3
  D. 4

_____ **16.** Based on a 50′ developed length of the vent pipe, a common bathtub with a 1¼″ individual vent has a dfu value of ___.
  A. 2
  B. 3
  C. 4
  D. 6

_____ **17.** A ___ group is a group of fixtures located next to a stack so vents may be reduced to a minimum using proper fittings.
  A. vertical
  B. horizontal
  C. stack
  D. vent

_____ **18.** Most municipalities maintain sewage disposal plants where the sewage treatment process consists of liquefying suspended ___ materials.
  A. organic
  B. inorganic
  C. hazardous
  D. all of the above

_____ **19.** ___ is relatively pure and can be discharged into a natural drainage basin without negatively affecting the ecology.
  A. Wastewater
  B. Waterborne waste
  C. Rainwater
  D. none of the above

_____ **20.** Building storm drains convey rainwater from roof drains through a rainwater ___ and into a storm sewer.
  A. router
  B. breaker
  C. levy
  D. leader

_____ **21.** When sizing stacks, no water closets are permitted to drain into a stack less than ___″ in diameter.
  A. 1
  B. 2
  C. 3
  D. 4

_____ **22.** One branch interval equals ___ floor(s) of plumbing fixture drains.
        A. one
        B. two
        C. three
        D. four

_____ **23.** Trap siphonage and back pressure are a result of pressure differences between the drainage system and ___.
        A. ventilation duct
        B. extension trap
        C. biosphere
        D. atmosphere

_____ **24.** Storm drain traps should be used when a roof drain body is located within ___′ of a door, window, or any other opening to a building.
        A. 4
        B. 6
        C. 8
        D. 10

_____ **25.** A(n) ___ vent is the most practical method of venting a fixture trap because trap seal loss is virtually eliminated.
        A. continuous
        B. individual
        C. relief
        D. group fixture

_____ **26.** Trap seal loss due to ___ pressure is prevented if fixture drains near the base of the stack or stack offset connect to the horizontal pipe at least 8′ from the offset.
        A. vent
        B. plus
        C. minus
        D. back

_____ **27.** A basic wet vent is a common vent with two fixtures whose horizontal drain openings are at different heights and installed ___.
        A. front to back
        B. side to side
        C. back to back
        D. front to front

_____ **28.** Vent terminals typically extend at least ___′ through the roof to prevent rainwater on the roof from draining into the terminals.
        A. 1
        B. 2
        C. 3
        D. 4

_____ **29.** Horizontal drainage piping is designed to flow about ___ full.
  A. one-third
  B. two-thirds
  C. three-fourths
  D. one-half

_____ **30.** In cold climates, vent pipes passing through the roof should be at least ___″ pipe.
  A. 1
  B. 2
  C. 3
  D. 4

_____ **31.** A yoke vent is connected to a soil stack with a wye and ___ bend.
  A. ⅛
  B. ¼
  C. ½
  D. ¾

_____ **32.** Not more than ___ dfu may drain into a 2½″ wet vent.
  A. 3
  B. 4
  C. 5
  D. 6

Completion

_____ **1.** A(n) ___ pipe is any pipe or fitting that makes an angle of less than 45° with the horizontal plane.

_____ **2.** ___ provides circulation of air to or from a sanitary drainage system and also provides air circulation within the sanitary and storm drainage piping.

_____ **3.** ___ convey only liquid waste that is free from fecal matter.

_____ **4.** Building sewers must be at least ___″ in diameter.

_____ **5.** A(n) ___ is any vertical line of soil, waste, or vent piping extending through one or more stories.

_____ **6.** A(n) ___ is a vertical length of stack at least 8′ high within which the horizontal branches from one story or floor of the building are connected to the stack.

_____ **7.** A(n) ___ drain is a drainage pipe extending horizontally from a soil or waste stack or building drain.

_____ 8. ___ is the length of vent pipe measured along the centerline of the pipe and fittings.

_____ 9. ___ refers to the various turns that may be required in drainage piping.

_____ 10. ___ is the slope of a horizontal run of pipe and is expressed as a fractional inch per foot length of pipe.

_____ 11. Sanitary drainage piping must be equipped with an adequate number of properly installed ___ to allow easy access to the system if stoppage occurs.

_____ 12. The ___ is a blanket of gases that surrounds the earth.

_____ 13. A(n) ___ is a combination of elbows or bends that brings one section of the pipe out of line but into a line parallel with the other section.

_____ 14. ___ pressure is a pressure less than 1 atmosphere.

_____ 15. Trap seal loss is a common problem in ___ piping.

_____ 16. A(n) ___ rim is the top edge of a fixture from which water overflows.

_____ 17. When a stack group is sized, the individual fixture drains are their normal size, but the stack vent is the same size as the ___ stack.

_____ 18. In most municipalities, private buildings are required to have drainage systems that connect to the ___ sewer.

_____ 19. In addition to the projected roof area, the size of the building storm drains is based on the horizontal storm drain ___.

_____ 20. Based on ___ tables, a plumber can establish the total discharge of all fixtures in a building in drainage fixture units and select a drain size to serve the demand.

_____ 21. A building drain should not be less than ___″ in diameter if it receives the discharge of three or more water closets.

_____ 22. A(n) ___ is any vertical line of soil, waste, or vent piping extending through one or more stories.

_____ 23. Hydrogen sulfide in a sanitary drainage system is objectionable because it absorbs additional oxygen from moisture in the drainage system, creating ___ acid, which is highly corrosive.

_____ 24. The size of the offset section of the stack is based on the ___ and number of drainage fixture units.

_____ 25. The ___ roof area is the area (in sq ft) of a portion of a roof drained by a particular pipe.

_____ 26. When planning the location of a building storm drain beneath a(n) ___ floor, the building storm drain and all branches are designed so that they will not cross over main runs of the building sanitary drain.

_____ **27.** Stacks that are 2″ and larger in diameter and that are over three stories or branch intervals high have ___ drainage fixture unit capacity than shorter stacks.

_____ **28.** Stack sizing varies depending on the ___ on which the fixtures are installed in a building.

_____ **29.** A yoke vent is connected to the waste stack below the horizontal fixture branch drain for that floor and to the vent stack at least ___ ″ above the floor level.

_____ **30.** In the colder climates, unwanted closure of the vent terminal due to ___ is a common problem and can cause trap seal loss.

_____ **31.** Storm water drain pipes are required to have cleanouts in the same locations as ___ pipes.

Matching

Venting Methods

_____ **1.** Vent stack

_____ **2.** Stack vent

_____ **3.** Yoke vent

_____ **4.** Relief vent

_____ **5.** Common vent

_____ **6.** Back vent

_____ **7.** Individual vent

_____ **8.** Wet vent

A. Extension of a soil or waste stack above the highest horizontal drain attached to the stack

B. Vent whose primary purpose is to provide additional air circulation between drainage and vent systems or to serve as an auxiliary vent on specially designed systems

C. Vent pipe provided specifically to prevent trap siphonage and back pressure

D. Vent pipe connecting upward from a soil or waste stack to a vent stack to prevent pressure differences in the stacks

E. Vent pipe that connects to a waste pipe on the sewer side of its trap to prevent siphonage

F. Vent pipe that vents a fixture trap and connects with the vent system above the fixture or terminates in the open air

G. Vent pipe that connects at the junction of two fixture drains and serves as a vent for both fixture drains

H. Vent pipe that is also a drain pipe

Roof Jackets and Flanges

_____ **1.** Warm climate

_____ **2.** Cold climate

_____ **3.** Flat

_____ **4.** Sloped

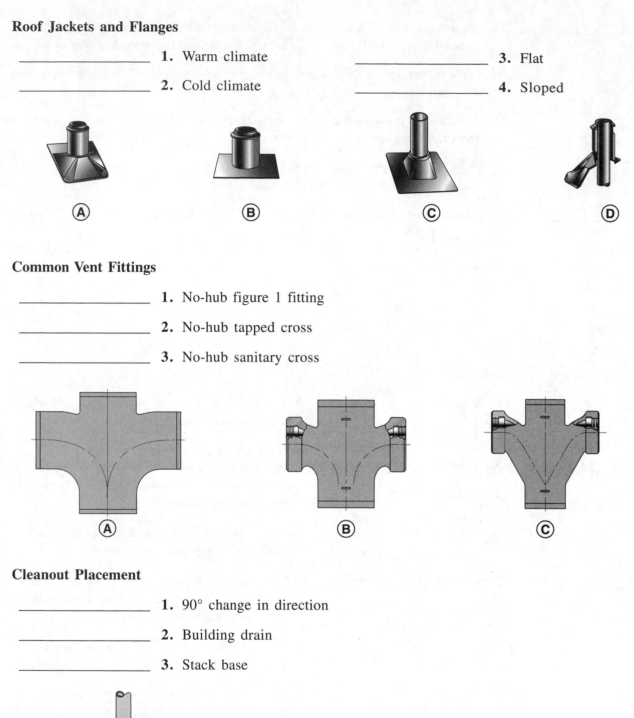

Common Vent Fittings

_____ **1.** No-hub figure 1 fitting

_____ **2.** No-hub tapped cross

_____ **3.** No-hub sanitary cross

Cleanout Placement

_____ **1.** 90° change in direction

_____ **2.** Building drain

_____ **3.** Stack base

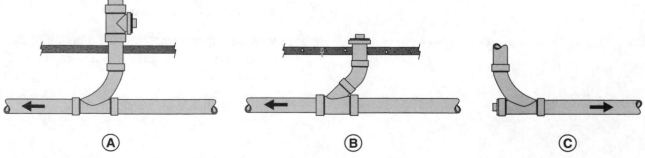

Short Answer

1. List five items a plumber must consider when sizing the sanitary drainage, vent, and storm water drainage systems of a building.

_____

_____

_____

_____

_____

2. Describe how the drainage fixture unit system was developed.

_____

_____

_____

_____

_____

_____

3. Explain why the efficiency of horizontal drainage pipe does not increase if a pipe larger than is necessary is installed.

_____

_____

_____

_____

4. Why is ¼″ per foot the recommended grade for horizontal drainage pipe?

_____

_____

_____

_____

5. What may be the result of installing horizontal drainage pipe at a grade greater than ¼″ per foot?

_____

_____

_____

_____

6. List five locations in which cleanouts should be provided.

_____

_____

_____

_____

_____

7. Distinguish between siphonage and back pressure.

_____

_____

_____

8. Why is pipe smaller than 1¼″ unsuitable for venting?

_____

_____

_____

_____

9. Why are individual vents commonly referred to as back vents?

_____

_____

_____

_____

_____

Activities

Sanitary Drainage and Vent Piping

Identify the sections of pipe shown on the drawing.

_____ **1.** Cleanout

_____ **2.** Branch

_____ **3.** Branch vent

_____ **4.** Roof terminal

_____ **5.** Waste pipe

_____ **6.** Soil stack

_____ **7.** Floor drain

_____ **8.** Common vent

_____ **9.** Building drain

_____ **10.** Horizontal branch

_____ **11.** Individual vent

_____ **12.** Stack vent

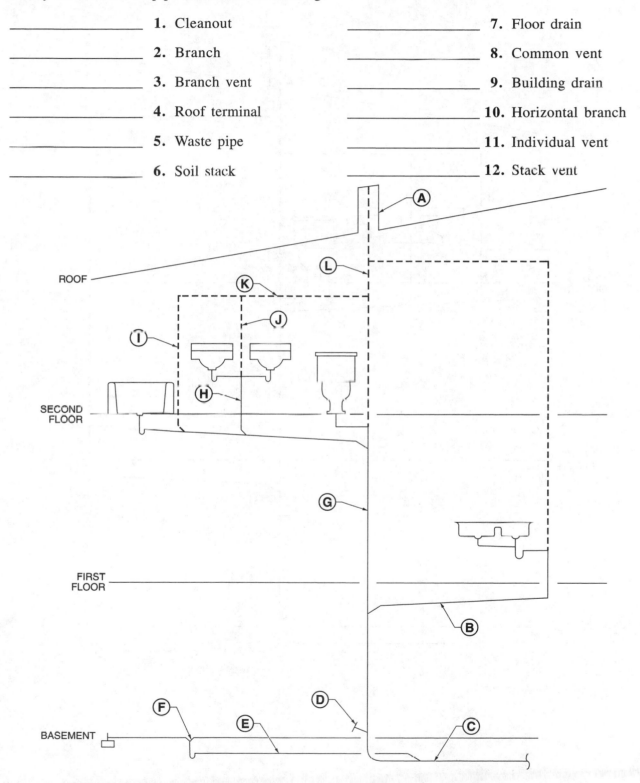

Sizing Sanitary Drainage and Vent Piping

1. Provide the minimum sizes of the entire sanitary drainage and vent piping system by referring to the tables in the textbook.

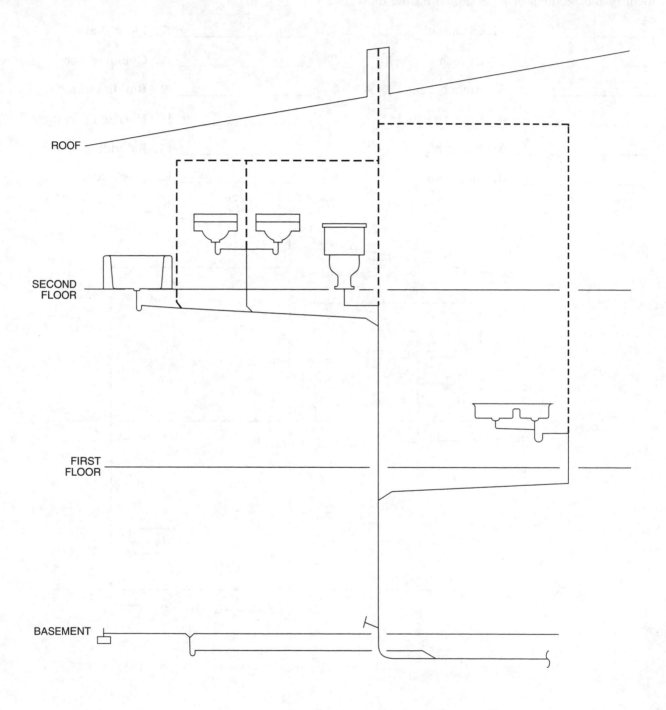

Drainage Fixture Units

1. Complete the drainage, waste, and vent system by adding lines where necessary. Give the minimum size of each section of the drainage, waste, and vent system. Indicate the dfu values for each fixture branch, stack, and entire system.

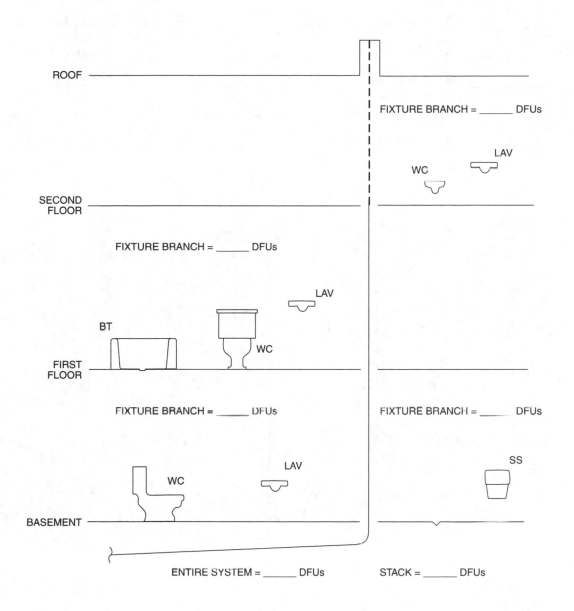

Sanitary Drainage and Vent Piping

Identify the sections of pipe, abbreviations, and symbols shown on the drawing.

_____ **1.** Individual vent

_____ **2.** Front main cleanout

_____ **3.** Water closet

_____ **4.** Stack vent

_____ **5.** Common vent

_____ **6.** Horizontal branch

_____ **7.** Bathtub

_____ **8.** Kitchen sink

_____ **9.** Roof terminal

_____ **10.** Lavatory

_____ **11.** P-trap

_____ **12.** Stack base cleanout

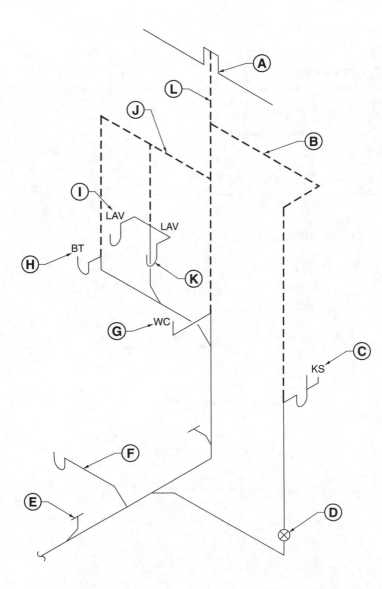

Sizing Sanitary Drainage and Vent Piping

1. Provide the minimum sizes of the entire sanitary drainage and vent piping system by referring to the tables in the textbook.

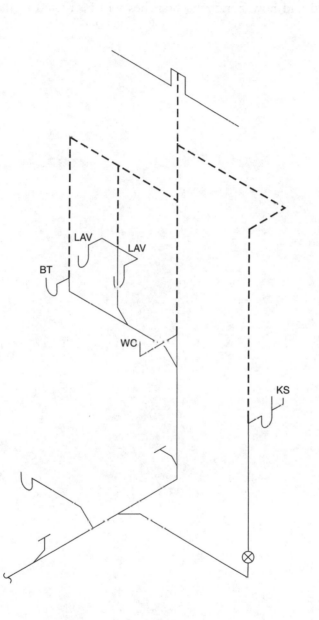

Projects

1. Sketch the sanitary drainage and vent piping system for your dwelling. Size the system on the local plumbing code.

2. Obtain a set of prints for a multifamily dwelling or small commercial building. Determine where the sanitary drainage piping enters the structure and the size of building sewer required. Identify where stacks will be installed and how horizontal branches will be routed to the stacks.

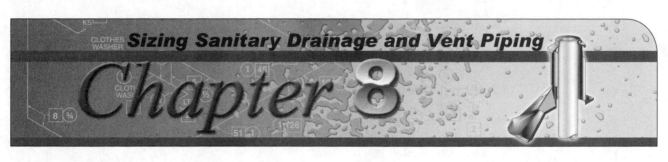
Name_____ Date _____

True-False

T	F	**1.** Drainage and venting loads are greater in commercial structures with a large number of occupants, making plumbing system design more challenging.
T	F	**2.** Minimum requirements are the absolute minimum level of piping that can be installed in a building.
T	F	**3.** A typical one-story, one-family dwelling consists of a bathroom and kitchen sink on the first floor, and a laundry tray and floor drain with 4″ trap in the basement.
T	F	**4.** In some jurisdictions, additional venting may be necessary to satisfy plumbing code requirements.
T	F	**5.** In a one-story, single-family dwelling with individual venting, all fixtures draining into the stack have a continuous waste and vent.
T	F	**6.** In a one-story, one-family dwelling with individual venting, a future vent is provided so there is a proper size vent opening if a basement bathroom is installed in the future.
T	F	**7.** Two-story, one-family dwellings typically include plumbing fixtures on both floors, as well as in a basement.
T	F	**8.** Drainage and vent pipe must be properly hung and supported to keep pipe properly aligned and to prevent sagging and leaking.
T	F	**9.** Vented closet crosses are available with back openings to accept the wastes of other fixtures.
T	F	**10.** A duplex residence includes different plumbing fixtures on each floor or area of the duplex.
T	F	**11.** In a typical duplex residence, individual vents need to be installed for the basement lavatory and for the bathtubs on the first and second floors.
T	F	**12.** Multifamily dwellings, such as apartments, have greater demands placed on the plumbing system than a one-family residence.
T	F	**13.** As a minimum requirement, two of the stacks in a typical multifamily dwelling serve the bathrooms for the four units.

T F **14.** In a two-story, one-family dwelling with individual venting, cleanouts are not included on the horizontal branches serving the second-floor bathtub and lavatory because there would be no access to the cleanouts in a residence.

T F **15.** In a multistory building, the bathroom waste and vent pipe connections are individually vented, similar to bathroom waste and vent piping in a residence.

T F **16.** In a two-story industrial building, gender-specific bathroom facilities must be provided.

Multiple Choice

_____ **1.** Based on the drainage fixture unit values of the fixtures, the minimum size(s) of the building ___ is(are) are determined.
  A. sewer
  B. drain
  C. drain branches
  D. all of the above

_____ **2.** In a one-story, one-family dwelling, the building drain, which is 3″ pipe sloped at ¼″ per foot, is increased to ___″ pipe for the building sewer, with the building sewer terminating at the sanitary sewer main or septic tank.
  A. 1
  B. 2
  C. 3
  D. 4

_____ **3.** A ___ is used to plumb the bathroom of the one-story, one-family dwelling, representing the minimum requirement for a waste and vent pipe installation.
  A. stack group
  B. vent stack
  C. roof vent
  D. all of the above

_____ **4.** In a one-story, one-family dwelling with individual venting, individual ___ are installed at the ends of the horizontal branches draining into the stack.
  A. trays
  B. cleanouts
  C. sinks
  D. waste inlets

_____ **5.** Two-story, one-family dwellings that are individually vented typically have water closets that drain into a ___″ sanitary tee and a wye in the stack.
  A. 1
  B. 2
  C. 3
  D. 4

_____ **6.** Waste from the lavatory on the first floor of a two-story, one-family dwelling with individually vented fixtures is drained by ___″ pipe into either a wye below the closet bend or into the side opening of the vented closet tee.
    A. 1
    B. 1¼
    C. 1½
    D. 2

_____ **7.** In a typical duplex residence, the laundry trays are connected to a 1½″ common waste pipe and are ___-vented with a 1¼″ pipe that is connected to the vent portion of the kitchen sink stacks.
    A. wet
    B. individual
    C. stack
    D. common

_____ **8.** In a typical duplex residence, the underground pipe serving the kitchen sink, laundry trays, and floor drain is ___″ pipe because it is a building drain branch from a stack.
    A. 1
    B. 2
    C. 3
    D. 4

_____ **9.** In a multifamily dwelling, common fixtures are installed ___.
    A. back-to-back
    B. side-to-side
    C. front-to-back
    D. front-to-front

_____ **10.** In a typical multifamily dwelling with four kitchen sinks, a 2″ pipe is used for the ___ portion of the waste stack because it receives 8 dfu of discharge.
    A. highest
    B. lowest
    C. largest
    D. strongest

Completion

_____ **1.** In a one-story, one-family dwelling, the front main cleanout extends ___″ minimum above the floor to prevent a homeowner or occupant from removing the cleanout plug and using the cleanout as a floor drain.

_____ **2.** For most plumbing codes, individually venting the bathtub, lavatory, and kitchen sink in a one-story, one-family dwelling is not necessary, provided that the distance between the trap and the stack does not exceed 2′–6″ for a ___″ lavatory trap.

_____   **3.** In a one-story, one-family dwelling with individual venting, the ___ branch drain that serves the lavatory is 1¼″ pipe.

_____   **4.** Two-story, one-family dwellings that are individually vented typically have a first-floor water closet set off to the side of the 3″ ___.

_____   **5.** A vented closet ___ and vented closet cross are specially designed fittings for offsetting a water closet on a lower floor from the stack to properly vent the water closet.

_____   **6.** Some multistory buildings, such as hotels, require only a ___ stack.

_____   **7.** In a multistory building that has more than six stories or branch intervals of bathrooms, a ___″ soil stack is required for the waste from the bathrooms.

_____   **8.** The bathtubs in a multistory bathroom stack may be wet vented through the ___ waste pipe.

_____   **9.** In the bathroom piping of a two-story office building, a 3 × 2 ___ is used to increase the vent pipe size to accommodate additional fixtures.

Matching

Water Closet Supporting Chair Carrier Fittings

_____   **1.** Horizontal branch

_____   **2.** Waste inlet

_____   **3.** Flushometer valve rough-in

_____   **4.** Vent

_____   **5.** Anchor foot

_____   **6.** Fixture bolt

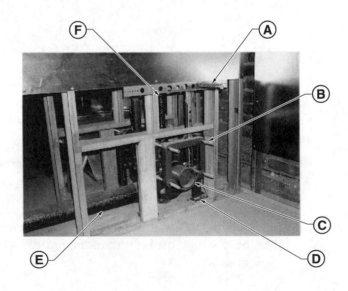

Short Answer

1. Explain why individual venting rather than stack group venting should be used in a structure.

_____

_____

_____

_____

_____

2. Discuss the sizing and installation problems that may occur in venting a system in a multistory structure.

_____

_____

_____

_____

_____

Activities

Sizing Sanitary Drainage and Vent Piping for One-Family Dwellings

Size each of the sections of pipe for a two-story, one-family dwelling according to local code.

_____ **1.** Size of building drain

_____ **2.** Size of building sewer

_____ **3.** Size of soil stack

_____ **4.** Size of vent above soil stack

_____ **5.** Size of bathtub trap

_____ **6.** Size of water closet vent

_____ **7.** Size of roof terminal

_____ **8.** Size of kitchen sink drain

_____ **9.** Size of laundry tray trap

Sizing Sanitary Drainage and Vent Piping for Multifamily Dwellings

Size each of the sections of pipe for a multifamily dwelling with back-to-back bathrooms according to local code.

_____ **1.** Size of bathtub traps

_____ **2.** Size of horizontal branches

_____ **3.** Size of common vents for bathtubs

_____ **4.** Size of soil stack

Sizing Sanitary Drainage and Vent Piping for Industrial Buildings

Size each of the sections of pipe for an industrial building according to local code.

_____ **1.** Size of drains for water closets

_____ **2.** Size of drains for urinals

_____ **3.** Size of drains for lavatories

_____ **4.** Size of vents for water closets

_____ **5.** Size of vents for urinals

_____ **6.** Size of soil stack

Sizing Sanitary Drainage and Vent Piping for Commercial Buildings

Size each of the sections of pipe for a multistory commercial building according to local code.

_____ **1.** Size of soil stack

_____ **2.** Size of soil branch for water closets

_____ **3.** Size of drains for lavatories

_____ **4.** Minimum size of vent stack

_____ **5.** Size of drains for urinals

Calculating Drainage Fixture Unit Values for a Two-Story, One-Family Dwelling

Use the isometric piping drawing to determine the dfu values for the building.

_____ **1.** dfu value for kitchen sink

_____ **2.** dfu value for floor drain

_____ **3.** dfu value for laundry tray

_____ **4.** dfu value for lavatory

_____ **5.** dfu value for water closet

_____ **6.** dfu value for bathtub

_____ **7.** Total dfu value for stack A

_____ **8.** Total dfu value for stack B

_____ **9.** Total dfu value for the system

10. Size the piping according to local code.

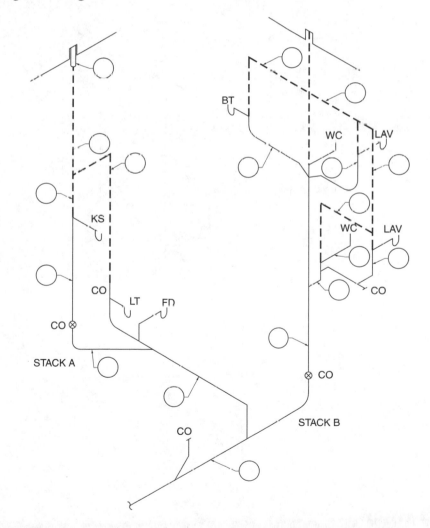

Calculating Drainage Fixture Unit Values for a Duplex Residence

Use the isometric piping drawing to determine the dfu values for the building.

_____ **1.** dfu value for water closet

_____ **2.** dfu value for laundry tray

_____ **3.** dfu value for kitchen sink

_____ **4.** dfu value for lavatory

_____ **5.** dfu value for bathtub

_____ **6.** Total dfu value for stack A

_____ **7.** Total dfu value for stack B

_____ **8.** Total dfu value for system

9. Size the piping according to local code.

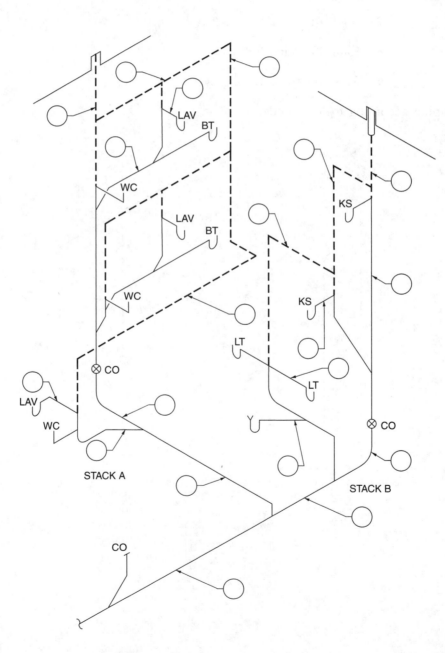

Sketching a Sanitary Drainage and Vent Piping System

1. Use the plan view to sketch an isometric piping drawing of the sanitary drainage and vent piping system for a one-story house with a basement using a minimum amount of venting. (Laundry facilities are in the basement.) Provide the minimum sizes of all sections of the piping.

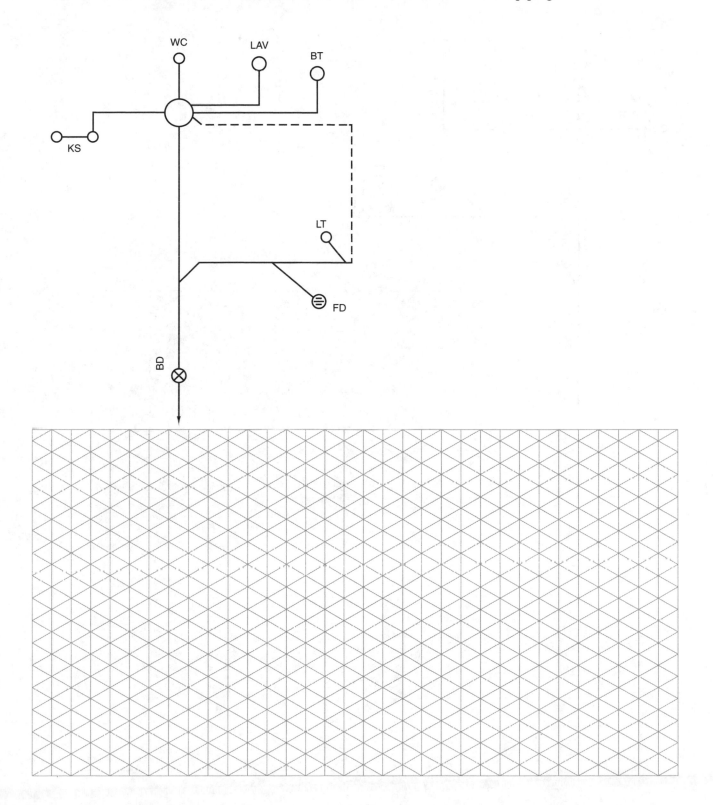

Sketching a Sanitary Drainage and Vent Piping System

2. Use the plan view to sketch an isometric piping drawing of the sanitary drainage and vent piping system for a one-story house with a basement. (Laundry facilities are in the basement.) Provide the minimum sizes of all sections of the piping.

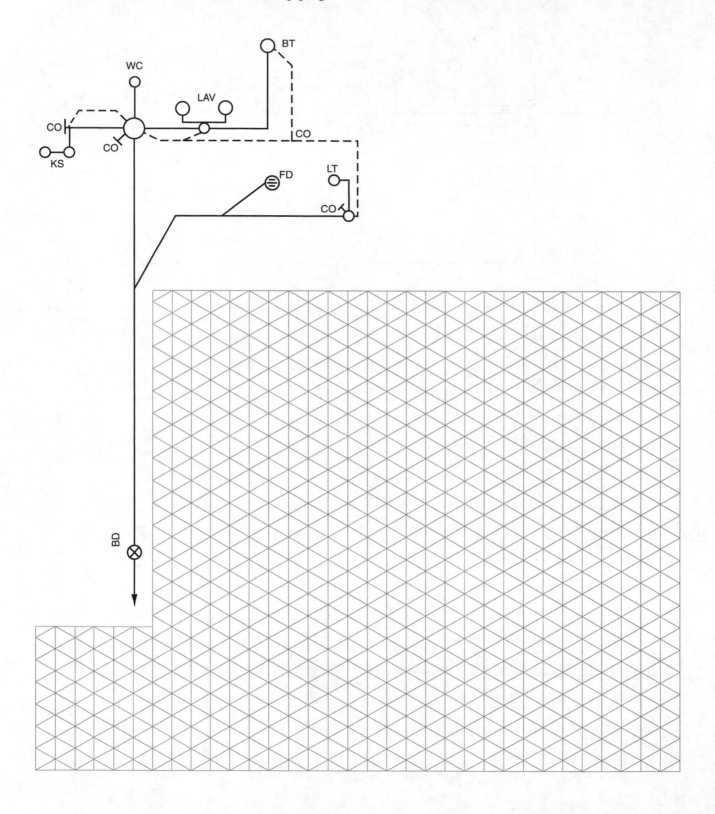

Sketching a Sanitary Drainage and Vent Piping System

3. Use the plan view to sketch an isometric piping drawing of the sanitary drainage and vent piping system for a two-story house with two bathrooms and a basement. (Laundry facilities are in the basement.) Provide the minimum sizes of all sections of the piping.

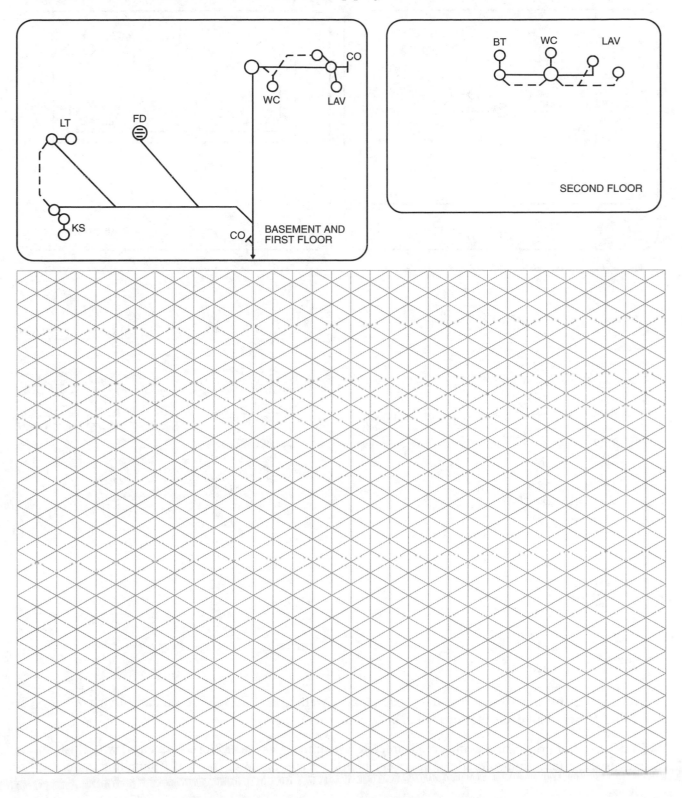

Sketching a Sanitary Drainage and Vent Piping System

4. Use the plan view to sketch an isometric piping drawing of the sanitary drainage and vent piping system for a duplex residence without a basement. Provide the minimum sizes of all sections of the piping.

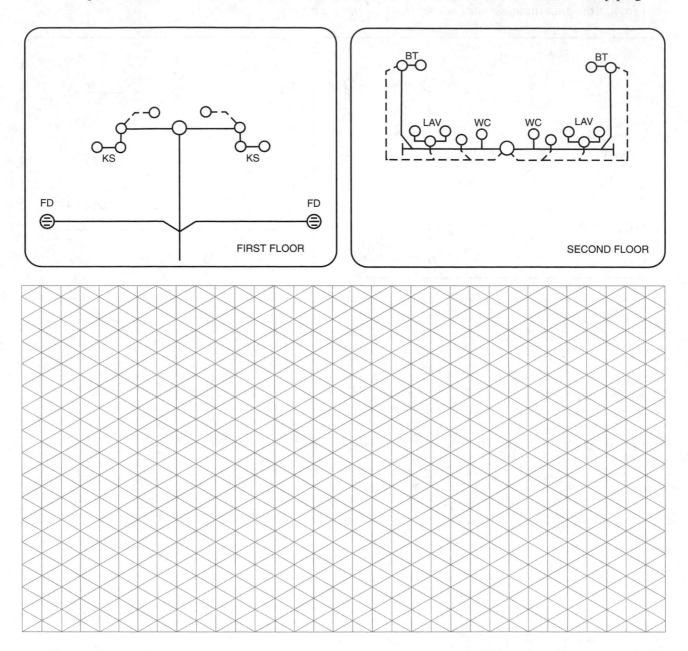

Projects

1. Obtain a set of prints for a one-story, one-family dwelling and design the sanitary drainage and vent piping system based on the local plumbing code.

2. Obtain a set of prints for a commercial structure which includes water closets equipped with flushometer valves. Size the sanitary drainage and vent piping system based on the local plumbing code.

Name_____ Date _____

True-False

T   F   **1.** Fixtures that do not have an integral trap must have a trap installed in the waste line as close to the fixture as possible.

T   F   **2.** An indirect waste pipe is a waste pipe that does not connect directly to the drainage system but conveys liquid waste by discharging into a plumbing fixture, interceptor, or receptacle that is directly connected to the drainage system.

T   F   **3.** Traps must have a smooth and uniform interior surface so waste does not accumulate and create a stoppage.

T   F   **4.** The additional resealing property of an anti-siphon P-trap is provided by a smaller volume of water retained in the trap.

T   F   **5.** One combination of fittings that conforms to trap seal depth and length requirements is a return bend and a 45° street elbow.

T   F   **6.** Oval-bottomed fixtures discharge their contents abruptly and do not allow the small amount of wastewater needed to reseal the trap to remain in the trap.

T   F   **7.** Self-siphonage may result from improper fixture trap venting.

T   F   **8.** When fixtures are not in use, the fixture drain pipe and living area where the fixtures are located are subjected to equal atmospheric pressure.

T   F   **9.** Fixtures in which back pressure occurs are usually located at the top of soil stacks or where soil pipe abruptly changes its direction.

T   F   **10.** The back pressure situation can be corrected by moving the water closet piping at least 8′ away from the base of the soil stack, and by properly venting the water closet.

T   F   **11.** Typically, the rate of evaporation is increased with lower temperatures.

T   F   **12.** Deep seal traps are recommended for fixtures that are used infrequently since deep-seal traps retain a greater volume of water used as a trap seal.

T   F   **13.** Water enters the trap through a primer tube connection in the P-trap or the floor drain body.

T   F   **14.** Traps must be protected from freezing, as freezing and thawing of the trap seal contents can crack or break traps allowing them to lose their trap seal contents.

T   F   **15.** While it is improbable that the entire trap seal will be displaced due to wind effects, precautions such as terminating the soil stack away from valleys, gables, or abrupt roof projections should be taken.

T   F   **16.** Trap leaks commonly occur under lavatory or kitchen sink cabinets.

T   F   **17.** Integral P-traps are used on water closets because the self-siphoning action of the trap is essential to the flushing action.

Multiple Choice

_____ **1.** A trap must be provided with a ___ or be designed so the trap can be easily disassembled.
   A. dip seal
   B. cleanout
   C. clean spout
   D. clean seal

_____ **2.** A(n) ___ , or ½ S-trap, is a type of trap with a dip for retaining wastewater in the dip to form a seal against sewer gases entering the living area.
   A. E-trap
   B. H-trap
   C. P-trap
   D. T-trap

_____ **3.** A trap ___ is the vertical distance between the crown weir and the top dip of the trap.
   A. seal
   B. inlet
   C. outlet
   D. dip

_____ **4.** A ___ seal P-trap is a P-trap with a 2″–4″ trap seal.
   A. clean
   B. half
   C. low
   D. common seal

_____ **5.** P-traps are available in 1¼″–6″ diameters and are constructed of various materials including ___.
   A. soil pipe
   B. brass
   C. DWV plastic
   D. all of the above

_____ **6.** To reduce the velocity of the waste entering the trap and the possibility of fouling, the maximum distance between the fixture outlet and the crown weir should not be greater than ___".
    A. 20
    B. 22
    C. 24
    D. 26

_____ **7.** ___ traps are typically used for specialized installations, such as in laboratories and dental offices, and for other fixtures that are difficult to properly vent, to prevent the loss of the trap seal by siphonage.
    A. Drum
    B. Dump
    C. No-hub
    D. Mechanically sealed

_____ **8.** Prohibited traps include ___.
    A. S-traps
    B. bell traps
    C. internal partition traps
    D. all of the above

_____ **9.** A ___ S-trap is an S-shaped plumbing trap with the outlet extending vertically downward from the crown weir.
    A. full
    B. half
    C. running
    D. bag

_____ **10.** A ___ trap is a prohibited trap in which the inlet and outlet of the trap are in vertical alignment with each other.
    A. gag
    B. bag
    C. slug
    D. key

_____ **11.** Trap seal loss is caused by ___.
    A. siphonage
    B. back pressure
    C. capillary attraction
    D. all of the above

_____ **12.** Plumbing codes also require that vents are not installed within ___ pipe diameters of the crown weir.
    A. one
    B. two
    C. three
    D. four

_____ **13.** A ___ vent is a vent pipe connected at the uppermost point in the crown of a trap.
   A. waste
   B. back
   C. crown
   D. head

_____ **14.** ___ is an objectionable occurrence in waste piping caused by plus pressure in which the trap seal contents and sewer gases are forced into the living area.
   A. Back pressure
   B. Back power
   C. Front pressure
   D. Loading pressure

_____ **15.** A type of trap primer is a(n) ___ trap primer.
   A. automatic
   B. water-saver
   C. vacuum breaker
   D. all of the above

_____ **16.** An automatic trap primer must be installed ___″ minimum above the top of the floor drain body and in a location that is accessible for maintenance and inspection.
   A. 6
   B. 8
   C. 10
   D. 12

_____ **17.** A ___ trap primer is a trap primer that uses the temporary head of water created when a lavatory is discharged to cause an overflow into the trap primer tube.
   A. head-saver
   B. water-saver
   C. surge tube
   D. primer valve

_____ **18.** A ___ breaker trap primer is a trap primer that is tapped into the side of the vacuum breaker tube on a water closet flushometer valve.
   A. flush
   B. circuit
   C. vacuum
   D. suction

_____ **19.** Capillary attraction is caused by a foreign object in the trap, such as a rag, string, lint, or hair, which extends over the ___ of the trap.
   A. inlet
   B. outlet
   C. base
   D. side

Completion

_____ **1.** A well-designed vent system is essential to maintaining a constant pressure of ___ atmospheres, which makes a trap effective.

_____ **2.** A(n) ___ is a fitting or device that, when properly vented, provides a liquid seal in drainage piping to prevent the escape of sewer gases into the living area.

_____ **3.** Each time a fixture is discharged, water is retained in the ___ to form a trap seal.

_____ **4.** A deep-seal P-trap is a P-trap with a trap seal over ___″.

_____ **5.** A(n) ___ P-trap is a P-trap with a large bowl on the outlet leg of a trap, and is designed to increase the resealing property of the trap.

_____ **6.** A(n) ___ trap is a plumbing trap in which the inlet and outlet are aligned horizontally, with a dip between the inlet and outlet to retain wastewater.

_____ **7.** A(n) ___ S-trap is a prohibited S-shaped plumbing trap with the outlet extending downward at an angle from the crown weir.

_____ **8.** Trap ___ is the process of conveying liquid through a pipe by means of a suction.

_____ **9.** ___ is a type of trap siphonage that is the result of unequal atmospheric pressures caused by rapid flow of wastewater through a trap.

_____ **10.** Siphonage by ___ results from a minus pressure in the waste piping caused by discharge of wastewater from an upper-level fixture into a fixture drain pipe that also serves a fixture installed at a lower level.

_____ **11.** ___ is a means of trap seal loss caused by the contents of the trap being absorbed into the surrounding air.

_____ **12.** A(n) ___ is a device that adds a small amount of water to the floor drain trap to ensure an adequate trap seal.

_____ **13.** A(n) ___ trap primer is a device with an integral air gap that is installed in a cold water supply pipe and diverts a small amount of water to a floor drain P-trap to ensure an adequate trap seal.

_____ **14.** The trap primer connection is located ___″ above the trap seal contents.

_____ **15.** ___ attraction is the upward movement of liquid, such as the contents of a trap seal, through fibers or cellular structure of a material.

_____ **16.** Minimum trap sizes are specified in ___ codes.

_____ **17.** P-traps must be installed as close to fixtures as possible to reduce fouling on the ___ side of the trap.

Matching

Prohibited Traps

_____ **1.** Internal partition

_____ **2.** Full S-trap

_____ **3.** S-trap

_____ **4.** Bell

_____ **5.** Mechanically sealed

_____ **6.** Bag

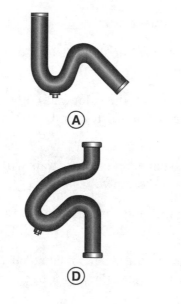

Ⓐ

Ⓑ

Ⓒ

Ⓓ

Ⓔ

Ⓕ

P-Traps

_____ **1.** Bottom dip

_____ **2.** Inlet

_____ **3.** Crown weir

_____ **4.** Crown

_____ **5.** Outlet

_____ **6.** Top dip

_____ **7.** 2″–4″ trap seal depth

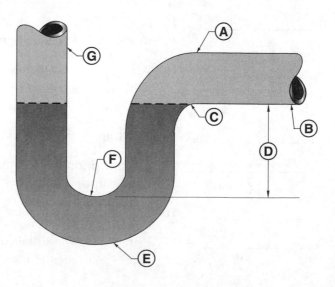

Automatic Trap Primer

_____ **1.** Primer tube connection

_____ **2.** Floor drain body

_____ **3.** Cold water supply pipe

_____ **4.** Automatic trap primer

_____ **5.** Air port

_____ **6.** Outlet

_____ **7.** Inlet

_____ **8.** Valve mechanism

Zurn Industries

Short Answer

List the minimum trap size for each of the following:

_____ **1.** Bathtub

_____ **2.** 2″ floor drain

_____ **3.** Lavatory

_____ **4.** Wall-hung urinal with 2″ trap

_____ **5.** Drinking fountain

_____ **6.** Service sink

_____ **7.** Bidet

8. Describe the conditions that may cause trap seal loss by evaporation.

_____

_____

_____

9. Explain what can be done to prevent trap seal loss by evaporation.

_____

_____

_____

10. One cause of trap seal loss is wind effect. High-velocity winds passing over the ridge of a roof and down the soil stack force water out of the trap. If this condition is encountered, what can be done to prevent trap seal loss due to wind effect?

_____

_____

_____

11. Describe self-siphonage and explain what can be done to prevent its occurrence.

_____

_____

_____

_____

12. Describe siphonage by momentum and what can be done to prevent its occurrence.

_____

_____

_____

_____

13. Describe back pressure and explain how back pressure can be prevented.

_____

_____

_____

_____

_____

14. Describe the operation of an automatic trap primer.

_____

_____

_____

_____

15. Describe the operation of a water-saver trap primer.

_____

_____

_____

_____

16. Describe the operation of a vacuum breaker trap primer.

_____

_____

_____

_____

17. List two disadvantages of deep-seal P-traps.

_____

_____

_____

Activities

Traps and Trap Seals

Sketch the following traps and shade the portion of the trap that is considered to be the trap seal.

1. P-trap   **2.** Drum trap   **3.** Running trap

Calculating Grade, Fall, and Run

Drainage piping grade calculations are frequently used on a plumbing job. Grade is the slope of a horizontal run of pipe and is expressed as a fractional inch per foot length of pipe. Fall is the distance the drainage pipe drops in its given length and is expressed in inches. Run is the horizontal distance the pipe covers and is expressed in feet.

Grade is calculated by dividing the fall by the run ($G = F \div R$). Fall is calculated by multiplying the run by the grade ($F = R \times G$). Run is calculated by dividing the fall by the grade ($R = F \div G$). Determine the unknown value for each of the following:

_____ **1.** Grade of a 200′ run of pipe that falls 50″

_____ **2.** Fall of a 75′ run of pipe that is graded at ⅛″ per foot

_____ **3.** Run of a pipe that falls 22″ and is graded at ¼″ per foot

_____ **4.** Fall of a 125′ run of pipe that is graded at ¼″ per foot

_____ **5.** Grade of a 50′ run of pipe that falls 12½″

_____ **6.** Run of a pipe that falls 45″ and is graded at ⅛″ per foot

_____ **7.** Fall of a 75′ run of pipe that is graded at ¼″ per foot

Projects

1. Obtain a set of residential prints and identify the fixtures and appliances to be installed. Determine the appropriate size traps for each of the fixtures or appliances.

2. Obtain a set of prints for a commercial building. Identify the fixtures and appliances to be installed in the building. Determine the types and sizes of traps to be used in the building.

Name_____ Date _____

True-False

T   F   **1.** At the point where a water heater fixture supply pipe joins a cold water main, the demand on the main increases to 329 cwsfu, yet it is not necessary to increase the cold water main size above the 2″ size originally selected, even though a larger size is indicated.

T   F   **2.** When street water main pressure fluctuates widely throughout the day, the water supply system of the building must be designed on the basis of the maximum pressure available.

T   F   **3.** Water pressure in a water supply system is increased due to the height to which the water must flow.

T   F   **4.** The total demand on a water supply system if all fixtures are used simultaneously is determined by adding the minimum flow rates for all the plumbing fixtures within a building.

T   F   **5.** A main is the principal pipe artery for a water supply, and fixture branches extend from the main to fixture supply pipes.

T   F   **6.** Head difference is a significant factor in a one-story building when determining proper sizing for water supply piping.

T   F   **7.** The available pressure is the water pressure in the street water main or other supply source.

T   F   **8.** Insufficient flow pressure results in excessive amounts of water flowing through the pipe and being delivered to the fixture.

T   F   **9.** Pressure loss due to friction increases as more pipe, fittings, valves, and other devices are installed in the water supply system.

T   F   **10.** Since water closets in a public building are fitted with flushometer valves, the cwsfu demand for the entire building is determined by adding the wsfu demand of the individual fixtures and assigning the appropriate values to the flushometer valves.

T   F   **11.** Water supply pipe size is based on the available water pressure, fixture demand, type and length of piping, height of the building, and flow pressure needed at the top floor.

T   F   **12.** Cold water pipe is sized beginning with the fixture nearest to the water meter.

T   F   **13.** Flow resistance varies with the pipe size and water flow rate.

Multiple Choice

_____   **1.** At the point where a ½″ lavatory fixture branch joins a ¾″ urinal fixture branch pipe, the demand increases to 14 cwsfu and the pipe increases to ___″ size.

     A. 1
     B. 1¼
     C. 1½
     D. 1¾

_____   **2.** Pressure loss due to head is subtracted from the ___ to determine whether there is adequate pressure to raise the water to the required height.

     A. total pressure
     B. available pressure
     C. wsfu
     D. cwsfu

_____   **3.** Water in street water mains is typically under a pressure of ___ psi, which is adequate to serve a moderate-size plumbing installation.

     A. 45 to 50
     B. 45 to 60
     C. 50 to 60
     D. 50 to 65

_____   **4.** A ¾″ ball valve is installed before the branch pipe passes through the dwelling wall so that water flow can be shut off ___.

     A. if the sillcock must be repaired
     B. during winter months to avoid freezing
     C. if the sillcock must be replaced
     D. all of the above

_____   **5.** Flow pressure ranges from 8 psi for faucets and tank-type water closets to ___ psi for certain models of flushometer valves.

     A. 18
     B. 22
     C. 25
     D. 27

_____   **6.** Water pressure within a building must never be allowed to exceed ___ psi.

     A. 60
     B. 70
     C. 80
     D. 90

_____ 7. When the cold water supply fixture unit demand increases to 51 cwsfu, the Water Supply Pipe and Water Meter Sizes for Water Supply Systems table indicates that the cold water main should increase to ___″ pipe at this point.
   A. 1
   B. 1¼
   C. 1½
   D. 2

_____ 8. Plumbing systems for residential structures are generally designed by ___.
   A. plumbers
   B. mechanical engineers
   C. structural engineers
   D. hydromechanical engineers

_____ 9. ___ is (are) used to convey water to fixtures.
   A. Galvanized pipe
   B. Copper tube
   C. PVC, CPVC, and PEX plastic pipe
   D. All of the above

_____ 10. The kitchen and service sinks, rated at 4 hwsfu each, are sized for ___″ fixture supply pipe.
   A. ¼
   B. ½
   C. ¾
   D. none of the above

_____ 11. The ___ is the only factor that can be affected by a plumber when installing a plumbing system.
   A. number of fixtures
   B. size of piping
   C. type of fittings
   D. length of piping

_____ 12. Flow pressure is determined by subtracting the pressure loss due to friction and pressure loss due to ___ from the available pressure.
   A. head
   B. leakage
   C. fixture use
   D. wear

_____ 13. The sillcock must be supplied by a ___″ branch pipe, as sized from the Minimum Sizes of Fixture Water Branch Pipe table.
   A. ¼
   B. ½
   C. ¾
   D. 1½

_____ **14.** The fixture supply pipe for women's rest room lavatories rated at 2 cwsfu each is ___" size.
　　　A. ¼
　　　B. ½
　　　C. ¾
　　　D. 1

_____ **15.** The water supply ___ is a measure of the estimated water demand of a plumbing fixture.
　　　A. estimate
　　　B. demand
　　　C. fixture unit
　　　D. rate

Completion

_____ **1.** ___ due to friction is the pressure variation resulting from friction within the pipe between the street water main and the water supply outlet where the water is being used.

_____ **2.** A(n) ___ valve is a device that lowers high and/or fluctuating water pressure to a lower and constant pressure appropriate for plumbing fixtures and appliances.

_____ **3.** The water supply fixture ___ method is a common method of properly sizing water supply piping for buildings that require a 2" or smaller water service and in which the distribution piping does not exceed 2½" size.

_____ **4.** Three fixtures that require a hot water supply—kitchen sink, lavatory, and bathtub—are properly served by ___" fixture branch and fixture supply pipes.

_____ **5.** As pressure loss within water supply piping increases, the discharge capacity of fixture supply pipes ___.

_____ **6.** The Minimum Sizes of Fixture Branch Pipe table indicates that a water closet with a flushometer valve has a ___" pipe inlet.

_____ **7.** ___ is the water demand by a fixture in a given amount of time, and is measured in gallons per minute (gpm).

_____ **8.** The ¾" water heater fixture branch pipe is fitted with a ¾" ___ valve before it connects to the water heater.

_____ **9.** ___, or working water pressure, is the water pressure in the water supply pipe near an outlet, such as a faucet, and is measured while the outlet is wide open and flowing.

_____ **10.** ___ and surface water constitute the sources of water used for residential, commercial, and industrial applications.

Matching

_____ **1.** Bathtub

_____ **2.** Relief valve

_____ **3.** Hot water

_____ **4.** Gate valve

_____ **5.** Service sink

_____ **6.** Ball valve

_____ **7.** Sillcock

_____ **8.** Water heater

_____ **9.** wsfu load

_____ **10.** Pipe size (in in.)

_____ **11.** Water closet

_____ **12.** Laundry tray

_____ **13.** Urinal

_____ **14.** Lavatory

_____ **15.** Water meter

_____ **16.** Cold water

_____ **17.** Kitchen sink

- - - -   (A)
———   (B)
½ (C)
2 (D)
B (E)
BT (F)

(G)
KS (H)
LT (I)
LAV (J)
RV (K)
SC (L)

SS (M)
UR (N)
WC (O)
(P)
WH (Q)

Short Answer

1. List three factors upon which the wsfu value for a fixture is based.

_____

_____

_____

2. List five factors that affect the size of water supply piping specified for a building.

_____

_____

_____

_____

_____

_____

3. Explain why the type of dwelling in which a fixture is installed must be considered when assigning a wsfu value.

_____

_____

_____

_____

_____

_____

4. Describe the relationship between water temperature and friction developed within water supply piping.

_____

_____

_____

_____

_____

5. When drilling holes in structural members of a building, why should the holes be drilled approximately ¼″ larger than the pipe size?

_____

_____

_____

Activities

Sizing Water Supply Piping

Size the following sections of water supply piping according to local code.

_____   **1.** First section of water pipe within a building

_____   **2.** Sillcock

_____   **3.** Unconcealed pipe for a drinking fountain

_____   **4.** Concealed pipe within a wall

_____   **5.** Pipe extending from the water main to water meter

_____   **6.** Fixture branch pipe

_____   **7.** Pipe for more than three fixtures or a branch

Wsfu Values for Flushometer Valves

Assign the wsfu value for each flushometer valve used in a large building.

_____ **1.** First flushometer valve

_____ **2.** Second flushometer valve

_____ **3.** Third flushometer valve

_____ **4.** Fourth flushometer valve

_____ **5.** Fifth flushometer valve

_____ **6.** Any additional flushometer valve

Sizing Water Supply Piping for a One-Story, One-Family Dwelling

Use the isometric piping drawing to calculate the demand for the water supply piping.

_____ **1.** Total wsfu = ___.

2. Size the piping based on the demand and according to local code.

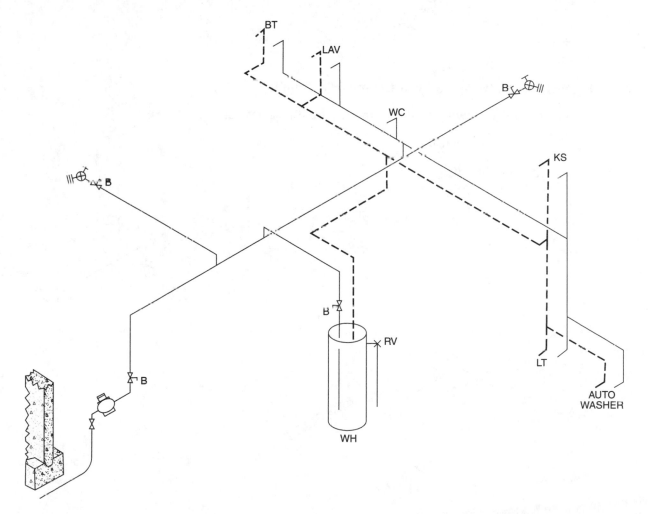

Sizing Cold Water Supply Piping for a Commercial Building

Use the isometric piping drawing to determine the cwsfu values for the building.

_____ **1.** Total cwsfu = ___.

2. Size the piping according to local code.

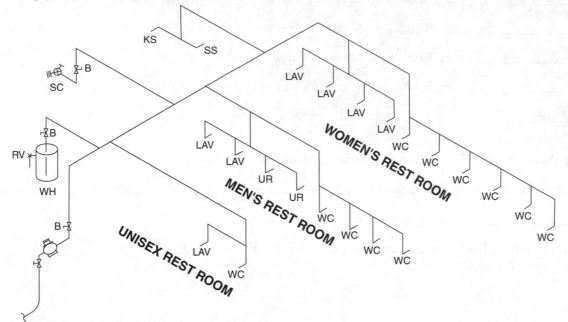

Sizing Hot Water Supply Piping for a Commercial Building

Use the isometric piping drawing to determine the hwsfu values for the building.

_____ **1.** Total hwsfu = ___.

2. Size the piping according to local code.

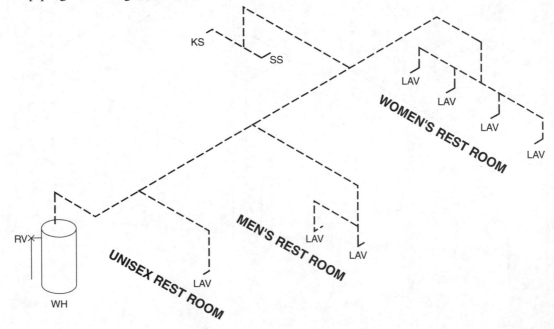

Sizing Cold Water Supply Piping for a Multifamily Dwelling

1. Size the cold water supply piping based on the supplied demand information and according to local code.

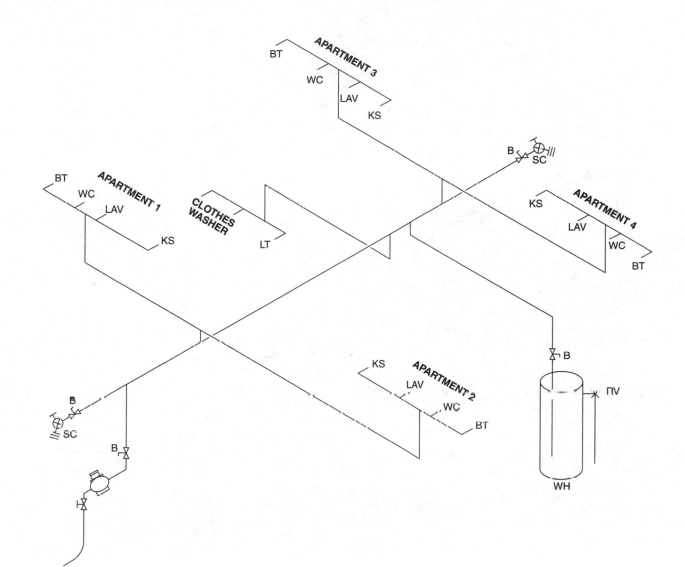

Sizing Hot Water Supply Piping for a Multifamily Dwelling

1. Size the hot water supply piping based on the supplied demand information and according to local code.

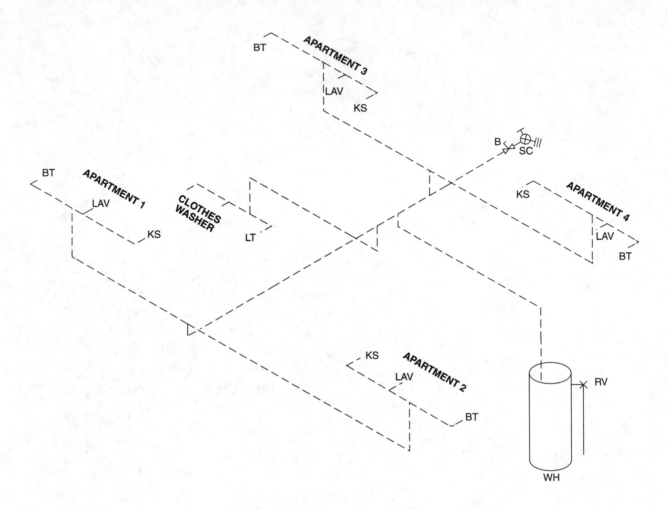

Designing Hot and Cold Water Distribution Systems for a Commercial Building

Use the DWV isometric piping drawing to calculate the cwsfu and hwsfu values for the building according to local code.

_____ **1.** Total cwsfu = ___

_____ **2.** Total hwsfu = ___

3. Draw and properly size the hot and cold water distribution system for the building. Use a blue pencil to indicate the cold water system and red to indicate the hot water system.

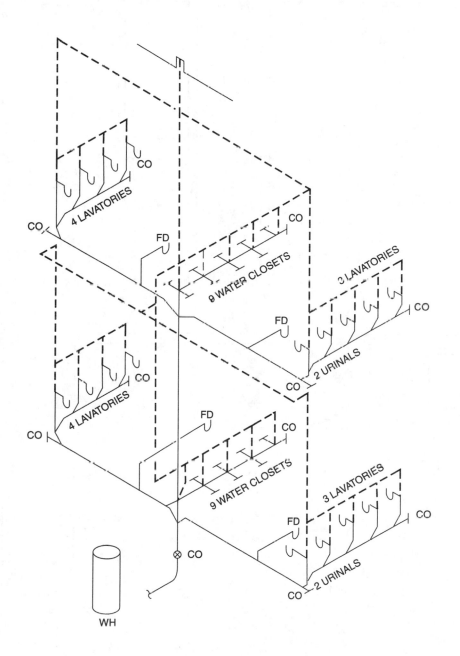

Designing Hot and Cold Water Distribution Systems for a Two-Story, One-Family Dwelling

Use the floor plan to calculate the total demand for the hot and cold water distribution system.

_____ **1.** Total cwsfu = ___

_____ **2.** Total hwsfu = ___

3. Draw an isometric piping drawing using a blue pencil to indicate the cold water system and red to indicate the hot water system. Properly size the isometric piping drawing according to local code.

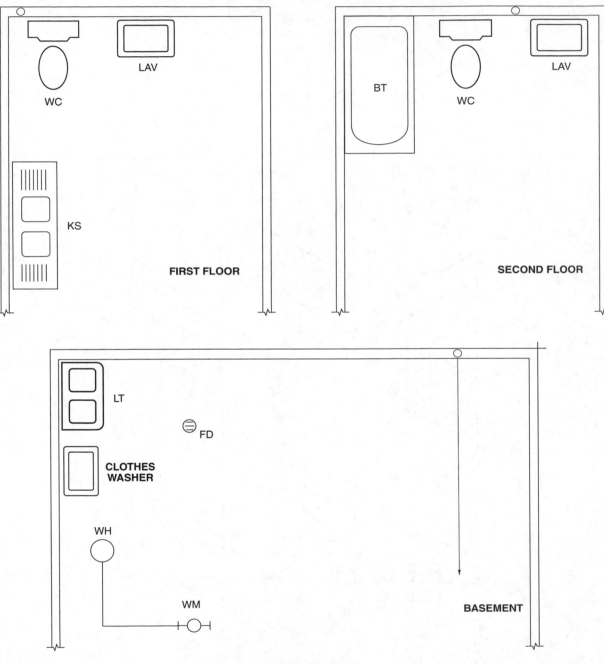

Designing Hot and Cold Water Distribution Systems for a One-Story, One-Family Dwelling

Use the floor plan to calculate the total demand for the hot and cold water distribution system.

_____ **1.** Total cwsfu = ___

_____ **2.** Total hwsfu = ___

3. Draw an isometric piping drawing using a blue pencil to indicate the cold water system and red to indicate the hot water system. Properly size the isometric piping drawing according to local code.

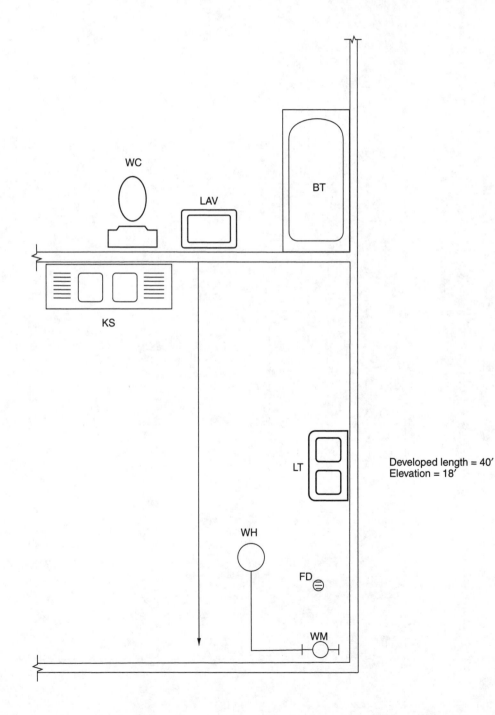

Developed length = 40′
Elevation = 18′

Projects

1. Sketch the water supply piping for your dwelling. Size the system according to local plumbing code.

2. Obtain a set of prints for a multifamily dwelling or small commercial building. Determine where the water supply pipe enters the structure and the size of water service required. Identify where water distribution pipes will be installed and how fixture branches will be routed to the fixtures and appliances.

Name_____ Date _____

True-False

T   F   **1.** If check valves fail and water supply pressure falls below the pressure within a boiler, boiler water and additives flow back into the potable water supply piping.

T   F   **2.** Hand-held tub and shower spray vacuum breakers are required by many plumbing codes if the spray hose is long enough for the spray head to extend past the flood level rim of the bathtub or lay on the shower stall floor.

T   F   **3.** A partial vacuum is a condition in which the pressure within a system is greater than atmospheric pressure.

T   F   **4.** A double check valve assembly (DCVA) is designed to be used under varying amounts of pressure and protect against back-siphonage only.

T   F   **5.** A barometric loop is a loop of water supply piping that rises 23′ and then falls back to the original piping elevation to protect against back-siphonage.

T   F   **6.** A hand-held tub and shower spray vacuum breaker is not required if the bathtub faucet operates the shower head through a diverter spout that falls back to the tub-fill position by gravity.

T   F   **7.** Potable water is water that is contaminated with impurities that could cause disease or harmful physiological effects.

T   F   **8.** Many plumbing codes require a separate permit be taken out for each reduced pressure zone (RPZ) backflow preventer device prior to installation so the utility department has a record of the device.

T   F   **9.** Backflow prevention devices must be protected from extremely hot or cold conditions.

T   F   **10.** Testing and rebuilding procedures for RPZ backflow preventers are performed by journeyman plumbers or apprentices and require no specialized training.

T   F   **11.** Double check valves with intermediate atmospheric vents are required for heating or cooling systems containing a glycol mixture.

T   F   **12.** A toxic substance is a solid, liquid, or gas that creates, or may create, danger to the health and well-being of consumers if it is introduced into the potable water supply.

T   F   **13.** An RPZ backflow preventer is a device that provides backflow protection for the potable water supply consisting of two independently operated check valves with an automatic operating differential relief valve located between the two check valves.

T   F   **14.** A PVB must be installed in a horizontal position.

T   F   **15.** CO_2 reacts with copper tube and forms a toxic compound that may cause illness or death.

T   F   **16.** Some backflow prevention devices are installed horizontally and require minimum distances between the flood level rim and the device.

Multiple Choice

_____ **1.** A(n) ___ is a type of backflow prevention device.
     A. atmospheric vacuum breaker
     B. double or dual check valve
     C. reduced pressure zone backflow preventer
     D. all of the above

_____ **2.** When pressure in a potable water supply system is below ___ psi, a partial vacuum can develop and cause back-siphonage.
     A. 10.8
     B. 12.3
     C. 14.7
     D. 15.2

_____ **3.** ___ is the flow of solids, liquids, or gases into the water distribution pipes of a potable water supply from any source other than the intended source.
     A. Backflow
     B. Crossflow
     C. Misflow
     D. none of the above

_____ **4.** A check valve ___.
     A. allows for testing of water pressure within a system
     B. admits air to the water supply through vents or ports to protect against back-siphonage
     C. relieves excess pressure from a potable water supply system
     D. permits fluid flow in only one direction and closes automatically to prevent backflow

_____ **5.** To prevent the float and disk from remaining (freezing) in the uppermost position, atmospheric vacuum breakers must be shut off at least once every ___ hr.
     A. 1 to 2
     B. 2 to 4
     C. 3 to 5
     D. 8 to 12

_____ **6.** A cross connection is any connection between the potable water supply system and ___ through which contaminants unfit for human consumption can enter the potable water supply pipe by back pressure or back-siphonage.
  A. a filtration system
  B. another piping system
  C. a fixture
  D. a reservoir or storage tank

_____ **7.** A ___ dual check valve is a device used to provide backflow protection for the potable water supply from cross connections within a residence.
  A. residential
  B. hose thread
  C. reduced pressure zone
  D. none of the above

_____ **8.** Plumbing codes require hose thread vacuum breakers to be installed on any hose thread connection except the drain valve of a ___.
  A. commercial water heater
  B. domestic water heater
  C. service sink
  D. sillcock

_____ **9.** A residential dual check valve consists of two replaceable check valves contained within a ___ or copper body.
  A. bronze
  B. galvanized steel
  C. cast iron
  D. none of the above

_____ **10.** For circular outlets, the air gap must be at least ___ the diameter of the potable water outlet and never less than 1″.
  A. one-half
  B. three-quarters
  C. twice
  D. three times

_____ **11.** A double check valve with intermediate atmospheric vent consists of two check valves within the body with a(n) ___ chamber between the valves.
  A. pressure-reducing
  B. atmospheric vent
  C. pressure-condensing
  D. reduced pressure zone

_____ **12.** An ABV consists of a float with a ___ disk which freely travels on a shaft within the body of the device.
  A. copper
  B. hard plastic
  C. rubber
  D. none of the above

_____ **13.** PVBs do not protect against ___.
   A. cross connection
   B. back-siphonage
   C. back pressure
   D. all of the above

_____ **14.** The ___ level is the level to which an AVB may be submerged before backflow through the device occurs.
   A. backflow
   B. overflow
   C. critical
   D. submersion

_____ **15.** ___ test cocks are installed on a double check valve assembly so the device can be tested to ensure the check valves are operating properly.
   A. Two
   B. Three
   C. Four
   D. Six

_____ **16.** The ___ is (are) responsible for enforcing the standards and regulations, as well as the supervision of public water supply systems and sources of drinking water.
   A. EPA
   B. states
   C. consumer
   D. supplier

_____ **17.** A ___ should be installed on the upstream side of an RPZ backflow preventer to minimize fouling of the check valves from debris flowing through the piping.
   A. strainer
   B. filter
   C. purifier
   D. all of the above

Completion

_____ **1.** ___ pressure refers to an installation in which pressure is continuously supplied to a backflow prevention device for over 12 hr.

_____ **2.** A(n) ___ substance is a substance that creates a non-health hazard or is a nuisance.

_____ 3. The ___ rim is the top edge of a fixture, such as a bathtub, from which water overflows.

_____ 4. Air gaps are used at the ends of water supply lines where reservoirs or ___ are installed, and they are integral to modern fixture design.

_____ 5. A(n) ___ is an opening through which water is conveyed to a fixture, boiler, or heating system; to equipment that is not part of the plumbing system but requires water to operate; or into the atmosphere.

_____ 6. The ___ is the diameter of a circle equivalent to the cross-sectional area of a potable water outlet opening.

_____ 7. Double check valves with intermediate atmospheric vents are typically installed in ___ systems to prevent water or other fluid used in the system from entering the potable water supply.

_____ 8. A(n) ___ is an unobstructed vertical distance through the air between the lowest outlet of any pipe or faucet supplying water to a plumbing fixture, tank, or other device and the flood level rim of the receptacle.

_____ 9. When water flows through a PVB, contents from downstream piping are vented to the ___ and air is introduced into the system to equalize the pressure.

_____ 10. A(n) ___ is a backflow prevention device used to protect against back-siphonage by admitting air to the water supply piping through vents or ports on the discharge side of the device.

_____ 11. Residential dual check valves are installed immediately downstream of the ___.

_____ 12. AVBs have a critical level, which is typically marked on the device with a ___.

_____ 13. The Safe ___ Act of 1974 established national standards and regulations for safe drinking water through the Environmental Protection Agency (EPA).

_____ 14. During normal flow condition, the check valves of a double check valve assembly are spring-loaded in the ___ position and require approximately 1 lb of water pressure to open and allow water passage.

_____ 15. Resilient-seated ball or ___ valves are installed at each end of an RPZ backflow preventer for water shutoff to the device for maintenance and repair.

_____ 16. RPZ backflow preventers must be rebuilt every ___ years, including the replacement of all internal rubber parts.

_____ 17. A beverage dispenser carbonator dual check valve consists of a body that contains three check valves—two ___ and one ball check valve.

Short Answer

1. Describe the operation of a hose thread vacuum breaker.

_____

_____

_____

_____

_____

2. Why do some municipalities require a residential dual check valve or double check valve assembly on residential water supplies?

_____

_____

_____

_____

3. List five conditions that cause an interruption in the potable water flow and affect the water supply pressure.

_____

_____

_____

_____

_____

4. What are the minimum clearance requirements for RPZ backflow preventers?

_____

_____

_____

_____

5. Identify two applications for an AVB.

_____

_____

_____

_____

6. What is the purpose of test cocks on PVBs?

_____

_____

_____

_____

7. Distinguish between back-siphonage and back pressure.

_____

_____

_____

_____

_____

8. Why must rust and other debris be prevented from entering a double check valve assembly during field testing?

_____

_____

_____

9. List four factors to consider when selecting and installing a backflow prevention device.

_____

_____

_____

_____

Matching

Backflow Prevention Devices

_____ 1. Atmospheric vacuum breaker

_____ 2. Pressure vacuum breaker

_____ 3. Hose thread vacuum breaker

_____ 4. Hand-held tub and shower spray vacuum breaker

_____ 5. Double check valve with intermediate atmospheric vent

_____ 6. Double check valve assembly

A. Small, inexpensive device with hose thread connections attached to sillcocks, service sinks, and other threaded water outlets where the potential exists for a hose being connected which could introduce a contaminant

B. Prevents back-siphonage of bathing water

C. Used for continuous or noncontinuous water pressure applications, and can be tested while installed in the piping

D. Protects the potable water supply in street water mains from backflow caused by cross connections within a building

E. Seals the atmospheric vent of the device under normal flow conditions and seals the device inlet when a partial vacuum develops in the water supply piping, allowing air into the system to break the vacuum

F. Used in continuous pressure applications to protect against back pressure and back-siphonage

Beverage Dispenser Carbonator Check Valve Operation

_____ 1. Normal flow condition

_____ 2. No flow condition

_____ 3. Back pressure condition

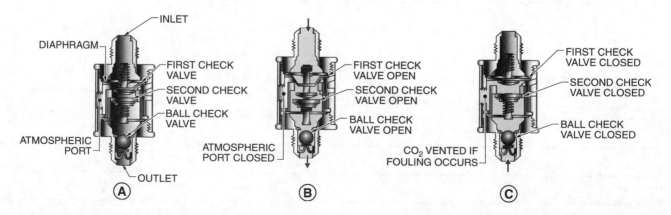

Residential Dual Check Valves

_____   **1.** Gate valve

_____   **2.** Residential dual check valve

_____   **3.** Water flow

_____   **4.** Ball valve

_____   **5.** Water meter

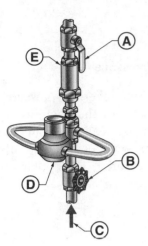

Atmospheric Vacuum Breaker

_____   **1.** Body

_____   **2.** Shaft

_____   **3.** Disk float

_____   **4.** Atmospheric vent

_____   **5.** Outlet

_____   **6.** Rubber disk

_____   **7.** Inlet

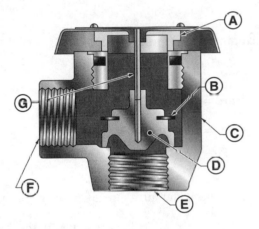

RPZ Backflow Preventer

_____   **1.** Vent port

_____   **2.** Inlet

_____   **3.** Outlet

_____   **4.** Second check valve

_____   **5.** Reduced pressure zone

_____   **6.** First check valve

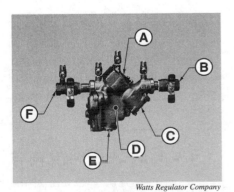

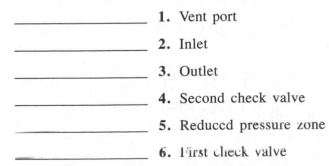

Watts Regulator Company

Activities

Calculating Pressure

A 1′ water column exerts a pressure of .434 pounds per square inch (psi) on its base, regardless of the shape or diameter of the container or pipe. Pressure is only affected by the height of the container or length of the pipe. Head is the amount of water pressure in a container or pipe at different elevations, and is expressed in feet. For example, a vertical section of 2½″ pipe measuring 3′ long has a 3′ head and a water heater measuring 18″ in diameter by 4′ in height has a 4′ head. The formula for calculating pressure when head is known is:

$$Pressure = Head \times .434$$

where

$Pressure$ = pressure (in psi)
$Head$ = head (in ft)
.434 = constant

For example, determine the pressure exerted by a head of 42′.

$Pressure = Head \times .434$

$Pressure = 42' \times .434$

$Pressure =$ **18.2 psi**

Determine the water pressure exerted at the base of the following containers or pipes.

_____ **1.** Pressure at the base of a 2″ × 27′ section of vertical ABS DWV plastic pipe

_____ **2.** Pressure at the base of a 15′ × 40′ cylindrical water storage tank

_____ **3.** Pressure at the base of a 4″ × 35′ section of no-hub cast iron soil pipe

_____ **4.** Pressure at the base of a 6″ × 48′ section of vertical PVC pipe

_____ **5.** Pressure at the base of a 1¼″ × 20′ section of vertical PVC pipe

_____ **6.** Pressure at the base of a 12″ × 120′ high section of Schedule 40 CPVC pipe

_____ **7.** Pressure at the base of a 2′ × 4′-9″ high water heater

_____ **8.** Pressure at the base of a ½″ × 10′ section of vertical PVC pipe

_____ **9.** Pressure at the base of a ¾″ × 18′ high section of type K copper tube

_____ **10.** Pressure at the base of a 6″ × 12′ high section of type L copper tube

Calculating Head

Each 1′ of head exerting .434 psi of water pressure on its base is equal to 2.3′ of head exerting 1 psi on its base. A pressure of 1 psi will force a quantity of water 2.3′ up a pipe. The formula for calculating head when pressure is known is:

Head = Pressure × 2.3

where

Head = head (in ft)
Pressure = pressure (in psi)
2.3 = constant

For example, determine the head produced with a pressure of 50 psi.

Head = Pressure × 2.3

Head = 50 psi × 2.3

Head = **115′**

Determine the head produced with the following water pressure.

_____ **1.** Pressure of 60 psi

_____ **2.** Pressure of 45 psi

_____ **3.** Pressure of 47 psi

_____ **4.** Pressure of 51 psi

_____ **5.** Pressure of 55 psi

Calculating Pressure Loss Due to Head

Water pressure in a water supply system is decreased due to the height to which water must flow. For every 1′ increase in elevation or head, there is a .434 psi loss of water pressure. For every floor of a building (10′), over 4 psi of water pressure is lost due to the height to which the water must flow. The formula for calculating pressure loss due to head is:

Pressure Loss Due to Head = Increase in Head × .434

where

Pressure Loss Due to Head = pressure loss due to height (in ft)
Increase in Head = increase in elevation of water level (in ft)
.434 = constant

For example, determine the pressure loss due to head when water is raised 30′.

Pressure Loss Due to Head = Increase in Head × .434

Pressure Loss Due to Head = 30′ × .434

Pressure Loss Due to Head = **13.02 psi**

Determine the pressure loss due to head for the following examples:

_____ **1.** 40′ rise in elevation

_____ **2.** Two story (20′) rise in elevation

_____ **3.** Rise in elevation from 15.1′ to 130.1′

_____ **4.** 68′ rise in elevation

_____ **5.** Rise in elevation from 46.8′ to 99.8′

Projects

1. Obtain backflow prevention device catalogs and price lists from a vendor such as Watts Regulator Company. Using the prints and specifications from a commercial construction project, identify backflow prevention devices that might be used for certain locations in the building, such as the entrance of the water service. Compare specifications, features, and prices for backflow prevention devices from several companies. Select a device for each location, considering the application and degree of hazard associated with the location.

Name_____ Date _____

True-False

T   F   **1.** A youth model water closet is a specialty wall-hung water closet.

T   F   **2.** A left hand bathtub is a bathtub with the drain on the right end as a person enters the tub.

T   F   **3.** Flush devices are designed to deliver the proper amount of water to the water closet bowl and provide the necessary scouring action to clean the bowl.

T   F   **4.** An integral temperature control device in port control valves is now available to maintain a safe water temperature and protect against a sudden change in water temperature, but plumbing codes do not require this.

T   F   **5.** Kitchen sinks require a basket strainer for each basin to prevent solid food particles from entering and clogging the drainage piping.

T   F   **6.** If the bypass port becomes plugged with dirt or other debris, a flushometer valve continues to flush until the dirt or debris is cleared since no water is allowed to enter the upper chamber.

T   F   **7.** A cast iron or plastic laundry tray drainage fitting is installed for a double-compartment laundry tray to keep the waste lines of both compartments separate before discharging into a 1½″ P-trap.

T   F   **8.** Standard wall-hung urinals are used as ADA-compliant urinals.

T   F   **9.** Fixture trim is the water supply and drainage fittings installed on a fixture or appliance to control water flow into a fixture and wastewater flow from the fixture to the sanitary drainage system.

T   F   **10.** A blowout water closet is a water closet that uses the gravity from water falling from the water closet tank to begin the flushing action.

T   F   **11.** A continuous waste fitting is a drainage fitting that consists of a section of horizontal drainage pipe and a sanitary tee and is used to convey waste from a kitchen sink drain to a common P-trap.

T   F   **12.** Vitreous china is a ceramic compound fired at high temperature to form a porous material which absorbs mineral deposits created by hard water.

T   F   **13.** Bathtubs are rated at 8 dfu.

T   F   **14.** A laundry tray, or laundry tub, is a plumbing fixture used for prewashing clothes installed in a residential laundry room, and is supplied with hot and cold water and a drain connection.

T   F   **15.** Blowout urinals have a large opening to the trap inlet to accommodate nonliquid waste.

T   F   **16.** A floor sink is a floor drain installed in commercial buildings to facilitate washing muddy boots or various types of cleaning and gardening equipment.

T   F   **17.** A wall-hung water closet is a water closet installed directly on the floor, and is common in residential construction.

T   F   **18.** Drop-in bathtubs typically provide greater water capacity than standard bathtubs.

T   F   **19.** The thermostat in an electric water heater cannot sense the outside surface temperature of the storage tank.

T   F   **20.** Most lavatory compression faucets are combination faucets in which the cold and hot water compression valves are joined in one faucet body and are provided with a common mixer spout.

T   F   **21.** Bi-level drinking fountains and water coolers are available to accommodate varying user heights and capabilities.

T   F   **22.** A drinking fountain is a wall-hung plumbing appliance that uses an electric cooling unit to chill drinking water, which is then conveyed through a nozzle at an upward angle to allow the user to conveniently drink from the fountain.

T   F   **23.** Soft water is potable water that contains excessive amounts of calcium and magnesium.

T   F   **24.** The dip tube of a water heater conveys incoming cold water through the stored hot water and discharges the cold water at the bottom of the storage tank.

T   F   **25.** Some plumbing codes do not require individual vents for floor drains if they are installed within 50′ of a vented drainage pipe.

T   F   **26.** An immersion element is an electric heating device that is inserted into the storage tank of an electric water heater and makes direct contact with the water to provide fast and efficient heat transfer to the water.

T   F   **27.** Raised-back, wall-hung lavatories support themselves better than ledge lavatories since there is more bearing surface on the back of the fixtures.

T   F   **28.** A T&P relief valve must be able to release high-temperature water at a rate equal to or faster than the rate at which the water heater can generate it.

T   F   **29.** Service sinks are provided with a grate fitting for the waste outlet and an enclosed P-trap.

T   F   **30.** An automatic water softener is typically piped into a residential water supply and provides soft water to all the plumbing fixtures and appliances except sillcocks and kitchen sink cold water faucets.

Multiple Choice

_____ **1.** A ___ faucet is a single-handle, noncompression faucet that contains ports for the hot and cold water supply and a cartridge or ceramic disc that opens and closes the ports as the faucet handle is moved or rotated.

     A. concealed

     B. port control

     C. cartridge

     D. centerset

_____ **2.** A ___ is a plumbing fixture that receives only liquid body waste and conveys the waste through a trap seal into a sanitary drainage system.

     A. water closet

     B. bidet

     C. lavatory

     D. urinal

_____ **3.** Water closets are rated at ___ dfu.

     A. 3

     B. 4

     C. 5

     D. 6

_____ **4.** When installing a water closet, water supply pipe is roughed in approximately 6″–8″ above the floor and is offset approximately ___″ to the left of the water closet centerline to provide easy access when servicing the flush tank.

     A. 4

     B. 6

     C. 6½

     D. 7

_____ **5.** A water-resistant ___ backing must be installed behind multipiece and ceramic tile showers.

     A. drywall

     B. plaster

     C. cement board

     D. masonry

_____ **6.** A water spot is ___.

     A. a stain left by the minerals found in hard water

     B. the surface area of the water in the water closet bowl when the bowl is empty

     C. the surface area of the water in the water closet bowl when a flush is completed

     D. an area of a building designated for the placement of water supply pipe

_____ **7.** ADA-compliant kitchen sinks feature a shallow bowl with the drain opening offset to the ___.
  A. front
  B. back
  C. left
  D. right

_____ **8.** A(n) ___ filter removes up to 99% of the undesirable elements in a water supply including lead, mercury, copper, pesticides, herbicides, nitrates, and tannins.
  A. carbon
  B. ionic
  C. reverse osmosis
  D. charcoal

_____ **9.** Plumbing codes require drinking fountains and water coolers to be manufactured with ___.
  A. fountain or cooler bowl construction of a nonabsorbent material
  B. a mouth protector provided over the nozzle
  C. the drinking water delivered at an angle so water cannot fall back onto the nozzle
  D. all of the above

_____ **10.** Common plumbing appliances include ___.
  A. water heaters
  B. water softeners
  C. water coolers
  D. all of the above

_____ **11.** A ___ automatic water softener is a water softener that has a mineral tank within the brine tank, and is the most common automatic water softener.
  A. two-tank
  B. closet
  C. cabinet
  D. single-tank

_____ **12.** A type of water closet is the ___ water closet.
  A. siphon jet
  B. gravity-fed
  C. blowout
  D. all of the above

_____ **13.** A(n) ___ flushometer valve is a flushometer valve in which a relief valve diaphragm within the valve body discharges water to a fixture by equalizing pressure on both sides of the valve.
  A. diaphragm
  B. electronic
  C. relief valve
  D. pressure-equalizing

_____ **14.** A ___ drain is a cast iron or plastic plumbing fixture set flush with the finished floor used to receive water drained from the floor and convey it into the drainage system.
    A. rainwater
    B. sewer
    C. floor
    D. none of the above

_____ **15.** Wall-hung urinal ___ support wall-hung urinals.
    A. hangers
    B. supports
    C. brackets
    D. none of the above

_____ **16.** A ___ is an electric appliance supplied with water, which grinds food waste into pulp and discharges the pulp into the drainage system.
    A. trash compacter
    B. food waste disposer
    C. juicer
    D. food grinder

_____ **17.** Water expands approximately ___% in volume for every 100°F of temperature increase.
    A. 1.25
    B. 2.5
    C. 2.75
    D. 3.25

_____ **18.** Siphon jet water closet bowls require ___ bolts to be properly anchored to the supporting chair carrier.
    A. two
    B. three
    C. four
    D. five

_____ **19.** Floor drains with integral P-traps are typically installed ___ ground level in buildings.
    A. at
    B. above
    C. below
    D. all of the above

_____ **20.** A ___ kitchen sink is one of the models available for installation.
    A. self-rimming
    B. metal-framed
    C. undercounter
    D. all of the above

_____ 21. Grab bars are required for ADA-compliant bathtubs on the ___ of a bathtub when the seat is in the fixture.
     A. foot end
     B. head end
     C. side wall
     D. all of the above

_____ 22. A ___, or shock absorber, is a device installed on water supply pipe near the fixture with the quick-closing valve to control the effects of water hammer.
     A. pressure buffer
     B. pressure-relief valve
     C. water hammer arrestor
     D. thermal expander

_____ 23. Combination waste and overflow fittings are manufactured from plastic or ___ tubing.
     A. copper
     B. brass
     C. cast iron
     D. steel

_____ 24. Some plumbing codes require installing a(n) ___ fitting in discharge piping from domestic dishwashers to prevent backflow of waste from a kitchen sink into the appliance.
     A. lift-and-turn waste
     B. gate valve
     C. air gap
     D. lift waste

_____ 25. A ___ handle is an ADA-compliant faucet handle that does not require tight grasping or twisting of the wrist to properly operate the faucet.
     A. wristblade
     B. wingblade
     C. wing knob
     D. flat knob

_____ 26. Rough-in includes installation of ___.
     A. drainage or vent piping
     B. water supply piping
     C. any necessary fixture supports
     D. all of the above

_____ 27. The bypass valve can be obtained when the water softener is purchased, or a bypass valve can be constructed with ___.
     A. two gate valves, one tee, and three short lengths of pipe
     B. two ball valves, two tees, and three short lengths of pipe
     C. three ball valves, two tees, and two short lengths of pipe
     D. two gate valves, one tee, and two short lengths of pipe

_____ **28.** A ___ is a floor-set basin used for cleaning and other maintenance tasks.
  A. floor sink
  B. utility sink
  C. maintenance basin
  D. mop basin

_____ **29.** Superheated water is water under pressure that is heated above ___°F without becoming steam.
  A. 212
  B. 225
  C. 282
  D. 312

_____ **30.** ___ is a synthetic resin bead used as the ion-exchange medium in the ion-exchange process to remove dissolved minerals from water.
  A. Polystyrene
  B. Zeolite
  C. Nylon
  D. Neoprene

Completion

_____ **1.** A(n) ___ is a plumbing fixture equipped with water and air circulation equipment and used to bathe and massage the entire body.

_____ **2.** A(n) ___ water closet is a water closet with a siphonic passageway at the rear of the bowl, and has an integral flush rim and jet.

_____ **3.** A(n) ___ is a self-closing, manually operated valve that delivers a pre-determined quantity of water to a washout urinal to remove the waste from the fixture.

_____ **4.** A(n) ___ is a plumbing fixture that performs a special function and is controlled and/or energized by motors, heating elements, or pressure- or temperature-sensing elements.

_____ **5.** A(n) ___ is a plumbing fixture used to bathe the external genitals and posterior parts of the body and also to provide relief for certain health conditions.

_____ **6.** A(n) ___, or toilet, is a water-flushed plumbing fixture that receives human liquid and solid waste in a water-containing receptacle and, upon flushing, conveys the waste to a soil pipe.

_____ **7.** Drinking fountain and water cooler water supply fittings have a self-closing ___ stop valve, which reduces the amount of wasted drinking water.

_____ **8.** A(n) ___ is a receptacle or device that is connected permanently or temporarily to the water distribution system, demands a supply of potable water, and discharges the waste directly or indirectly into the sanitary drainage system.

_____ **9.** ___ is a water supply system defect in which a loud noise is created when a quick-closing valve, such as a clothes washer valve or ball valve, is suddenly closed.

_____ **10.** A washout urinal has a restricted opening over the trap inlet consisting of a stainless steel strainer, ___, or small openings cast into the fixture to prevent debris from entering the trap and plugging the fixture drain.

_____ **11.** A(n) ___ is a flush device actuated by direct water pressure to supply a fixed quantity of water for flushing purposes.

_____ **12.** A(n) ___ is a shallow, flat-bottomed plumbing fixture that is used to clean dishes and prepare food.

_____ **13.** A(n) ___ is an orifice (opening) at the base of the bowl that directs water into the passageway inlet to help create the siphonic flushing action.

_____ **14.** A(n) ___ is a plumbing fixture used to wash the hands and face.

_____ **15.** ___ is (are) installed at the back and side wall of an ADA-compliant water closet to aid a user in transferring from a wheelchair to the fixture.

_____ **16.** A(n) ___ is a plumbing fixture used to bathe the entire body.

_____ **17.** A(n) ___ is a wall-hung plumbing appliance that incorporates an electric cooling unit into a drinking fountain to provide cooled drinking water at a desired temperature.

_____ **18.** A(n) ___ is a drain fitting installed in a kitchen sink that consists of a strainer body fitted with a fixed strainer and a removable basket with a rubber stopper.

_____ **19.** A(n) ___ is a plumbing fixture that discharges water from above a person who is bathing.

_____ **20.** Compression or ___ faucets are installed in showers to control the volume and temperature of water flowing from the shower head.

_____ **21.** ___ are individual atoms or groups of atoms that carry an electrical charge.

_____ **22.** A(n) ___ faucet is a faucet in which the flow of water is shut off by means of a washer that is forced down onto its seat, as in a globe valve.

_____ **23.** A(n) ___ is a small electric generator made of two different metals that are firmly joined, and is used as a safety device for gas-powered appliances.

_____ **24.** ADA-compliant bathtubs are equipped with a(n) ___ and grab bars to assist the user in entering and exiting the fixture.

_____ **25.** A(n) ___ is a plumbing appliance that removes dissolved minerals, such as calcium and magnesium, from water by an ion-exchange process.

_____ **26.** A clothes washer ___ is a plastic enclosure that accommodates water supply and waste connections for a clothes washer.

_____ **27.** ___ is the installation of all parts of a plumbing system that can be completed prior to the installation of the fixtures.

_____ **28.** A(n) ___ is an electric plumbing appliance used to wash dishes.

_____ **29.** ___ is the final installation, or setting, of plumbing fixtures and appliances.

_____ **30.** A(n) ___ is a plumbing appliance that removes sand, sediment, chlorine, lead, and other undesirable elements from water and protects water heaters and other fixtures and appliances from collecting residue.

_____ **31.** A(n) ___, or slop sink, is a high-back sink with a deep basin used for filling and emptying scrub pails, rinsing mops, and disposing of cleaning water.

_____ **32.** A(n) ___ relief valve is an automatic self-closing safety valve installed in the opening in a water heater that protects against the development of high temperature and/or high pressure within the storage tank.

_____ **33.** A(n) ___ is a plumbing appliance used to heat water for purposes other than heating a structure.

_____ **34.** A(n) ___ is a reservoir that retains a supply of water used to flush one water closet.

Short Answer

1. Explain the purpose of water filters in a water supply system.

_____

_____

_____

_____

2. Describe the operation of a ceramic disk-type port control faucet.

_____

_____

_____

_____

3. Explain how a water closet passageway is measured.

_____

_____

4. Describe the purpose of the Americans with Disabilities Act and how its requirements affect installation of fixtures and appliances.

_____

_____

_____

_____

_____

_____

5. Why are blowout water closets commonly installed in healthcare facilities?

_____

_____

6. How do urinal flushometer valves differ from water closet flushometer valves?

_____

_____

_____

_____

7. Describe the operation of an electronic flushometer valve.

_____

_____

_____

_____

_____

8. Explain the differences between pressure-balancing and thermostatic valve shower faucets.

_____

_____

_____

9. Describe the preparations required prior to installing a shower base.

_____

_____

_____

_____

10. List the procedure for mounting a food waste disposer below a kitchen sink basin.

_____

_____

_____

_____

_____

_____

11. Describe the operation of a hand valve.

_____

_____

_____

12. Explain the purpose of a floor sink in a commercial kitchen.

_____

_____

_____

13. List four typical locations of floor drains.

_____

_____

14. Describe the operation of a flush tank.

_____

_____

_____

_____

_____

_____

_____

15. List five typical effects of hard water.

_____

_____

_____

_____

_____

_____

_____

16. Describe the ion-exchange process of water softeners.

_____

_____

_____

_____

_____

_____

_____

Matching

Faucets and Bathtubs

_____ 1. Port control faucet

_____ 2. Freestanding bathtub

_____ 3. Centerset faucet

_____ 4. Recessed bathtub

_____ 5. Drop-in bathtub

_____ 6. Concealed faucet

A. A bathtub permanently attached to wall studs or backing boards behind the fixture and supported by the floor

B. Combination lavatory fitting that consists of one or two faucet handles and a spout mounted above the lavatory or countertop, and a pop-up waste fitting lift rod with the faucet bodies below the fixture

C. Combination lavatory fitting that consists of a single-handle faucet, spout, and pop-up waste fitting lift rod mounted on a raised base

D. Combination lavatory fitting that consists of two faucet handles, a spout, and a pop-up waste fitting lift rod mounted on a raised base

E. Bathtub that is installed in an enclosure that supports the fixture

F. Bathtub supported by legs and not permanently attached to the bathroom walls or floor

Bathtub Fittings

_____ 1. Combination waste and overflow fitting

_____ 2. Overrim bathtub fitting

_____ 3. Lift-and-turn waste fitting

_____ 4. Lift waste fitting

A. A bathtub water supply fitting that consists of a faucet assembly and mixing spout

B. A combination waste and overflow fitting consisting of a stopper with a raised knob at the top that is raised and rotated to allow the fitting to remain in the drain position

C. A combination waste and overflow fitting in which a lifting mechanism within the overflow tube is connected by a lever to the stopper in the bathtub drain outlet

D. A bathtub drain fitting that is an outlet for bathtub waste and allows excess water to drain from the fixture so it does not overflow onto the bathroom floor

Urinals and Lavatories

_____ 1. Washout urinal

_____ 2. Wall-hung lavatory

_____ 3. Countertop lavatory

_____ 4. Self-rimming lavatory

_____ 5. Siphon jet urinal

_____ 6. Pedestal lavatory

_____ 7. Vanity-top lavatory

_____ 8. Blowout urinal

_____ 9. Undercounter lavatory

A. A two-piece lavatory with the wash basin resting directly on a pedestal base

B. A lavatory supported by a stamped steel or cast iron bracket fastened to a backing board installed between wall framing members when the fixture is roughed in

C. A urinal in which the trap seal is washed from a trap through a restricted opening at the trap inlet

D. A lavatory installed in an opening of a bathroom or rest room cabinet or countertop, or resting on a cabinet frame

E. A wall-hung urinal in which the trap seal is forced from a trap through a large opening at the trap inlet

F. A lavatory attached to the underside of a countertop using rim clamps or other proprietary hardware

G. A wall-hung urinal with a nonsiphonic passageway at the rear of the bowl and an integral flush rim and jet

H. A one-piece wash basin and countertop installed on top of a bathroom or rest room cabinet or supported by wall framing members

I. A lavatory in which the bowl is placed in an opening in the countertop, and the edge of the fixture rests directly on top of the countertop

Blowout Water Closet Operation

_____ 1. Priming jet

_____ 2. Flush rim

_____ 3. Trap seal

_____ 4. Flushometer valve connection

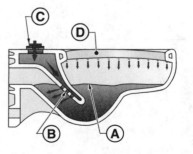

Water Closet Flush Tanks

_____ 1. Water control float

_____ 2. Flush valve assembly

_____ 3. Water control
assembly

_____ 4. Rubber flapper

_____ 5. Handle

_____ 6. Refill tube

_____ 7. Trip lever

_____ 8. Water inlet

_____ 9. Water outlet

_____ 10. Lift chain

_____ 11. Overflow tube

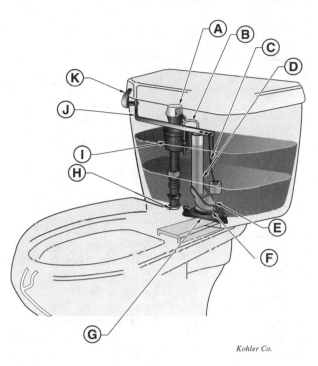

Kohler Co.

Food Waste Disposers

_____ 1. Internal baffle

_____ 2. Continuous waste
fitting

_____ 3. Mounting ring

_____ 4. Waste outlet

_____ 5. Sink flange

_____ 6. Baffle tee

_____ 7. Dishwasher drain
connection

_____ 8. Basket strainer

_____ 9. Food waste disposer

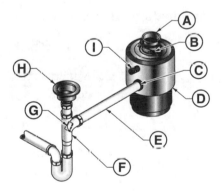

Diaphragm Flushometer Valves

_____ 1. Control stop

_____ 2. Outlet

_____ 3. Upper chamber

_____ 4. Diaphragm

_____ 5. Lower chamber

_____ 6. Plunger

_____ 7. Handle

_____ 8. Relief valve

_____ 9. Bypass port

_____ 10. Valve body

_____ 11. Handle coupling

_____ 12. Inlet

_____ 13. Tailpiece

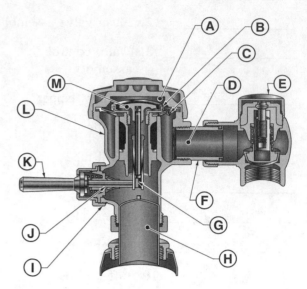

Cabinet Automatic Water Softeners

_____ 1. Storage tank

_____ 2. Mineral tank with zeolite

_____ 3. Brine valve

_____ 4. Multi-port valve assembly

_____ 5. Demand timer control

_____ 6. Sodium storage tank cover

_____ 7. Faceplate

_____ 8. Brine well

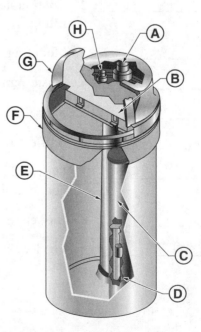

Activities

Fixture and Appliance Minimum Requirements

Per local plumbing code, list the minimum requirements for the following fixtures and appliances.

Water Closet

_____ **1.** Minimum requirement for vent

_____ **2.** Minimum requirement for waste

_____ **3.** dfu value

_____ **4.** Type of faucet/valve flush tank

_____ **5.** Type of trap

Wall-Hung Urinal

_____ **1.** Minimum requirement for vent

_____ **2.** Minimum requirement for waste

_____ **3.** dfu value

_____ **4.** Type of faucet/valve

_____ **5.** Type of trap

Kohler Co.

Lavatory

_____ **1.** Minimum requirement for vent

_____ **2.** Minimum requirement for waste

_____ **3.** dfu value

_____ **4.** Type of faucet/valve

_____ **5.** Type of trap

Bathtub

_____   **1.** Minimum requirement for vent

_____   **2.** Minimum requirement for waste

_____   **3.** dfu value

_____   **4.** Type of faucet/valve is

_____   **5.** Type of trap

Kohler Co.

Shower

_____   **1.** Minimum requirement for vent

_____   **2.** Minimum requirement for waste

_____   **3.** dfu value

_____   **4.** Type of faucet/valve

_____   **5.** Type of trap

Kohler Co.

Drinking Fountain

_____   **1.** Minimum requirement for vent.

_____   **2.** Minimum requirement for waste

_____   **3.** dfu value

_____   **4.** Type of faucet/valve

_____   **5.** Type of trap

Elkay Mfg. Co.

Kitchen Sink

_____ 1. Minimum requirement for vent

_____ 2. Minimum requirement for waste

_____ 3. dfu value

_____ 4. Type of faucet/valve

_____ 5. Type of trap

Kohler Co.

Laundry Tray

_____ 1. Minimum requirement for vent

_____ 2. Minimum requirement for waste

_____ 3. dfu value

_____ 4. Type of faucet/valve

_____ 5. Type of trap

Service Sink

_____ 1. Minimum requirement for vent

_____ 2. Minimum requirement for waste

_____ 3. dfu value

_____ 4. Type of faucet/valve

_____ 5. Type of trap

Projects

1. Obtain a set of residential prints and identify the fixtures and appliances required for the residence. Develop a list of the fixtures and appliances and obtain model numbers from manufacturer catalogs or Web sites for the fixtures and appliances.

2. Obtain a set of prints for a commercial building and identify the fixtures and appliances required for the building. Develop a list of the fixtures and appliances and obtain model numbers from manufacturer catalogs or Web sites. Determine the amount of time required to rough-in and install the fixtures and appliances.

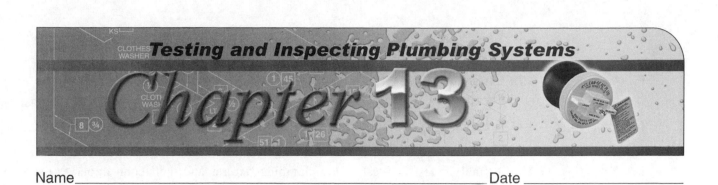
Name_____ Date _____

True-False

T   F   **1.** Testing plumbing systems is the responsibility of the plumber.

T   F   **2.** For some installations, several plumbing systems can be tested and inspected simultaneously, such as the building sewer and water service when they are installed in the same trench.

T   F   **3.** Hydrostatic tests are used on plumbing systems constructed of metallic piping, while air or hydrostatic tests are typically used on plumbing systems constructed of plastic piping.

T   F   **4.** An air test is a plumbing system test in which inlets and outlets to the system are sealed and air is forced into the system until a uniform air pressure of 5 psi is reached and maintained for 15 min without additional air being added to the system.

T   F   **5.** A hand pump or portable air compressor is used to force air into the plumbing system, and is typically used when testing smaller installations, such as a one-family dwelling.

T   F   **6.** Air pockets in the water service piping must be eliminated prior to testing.

T   F   **7.** Always wear a protective helmet (hard hat) and eye protection when performing air and hydrostatic tests using test plugs and do not stand directly in front of an outlet where a test plug is installed.

T   F   **8.** Hydrostatic tests are typically performed on PEX systems since it is less time-consuming to perform a hydrostatic test than an air test.

T   F   **9.** Underground sanitary drainage and vent piping is tested and inspected after it is covered with earth and backfilled.

T   F   **10.** A sanitary drainage and vent piping air test is performed by attaching an air compressor to a suitable opening and closing all other inlets and outlets to the system with pipe caps, pipe plugs, or test plugs.

T   F   **11.** In tall buildings, the sanitary drainage and vent piping is typically tested in sections because of the tremendous pressure produced by the water column height if the entire height of the pipe is filled with water.

T	F	**12.**	Water in the sanitary drainage and vent piping is kept in the system upon completion of a hydrostatic test.
T	F	**13.**	Exterior rainwater leaders that do not connect to the storm sewer system are not required by plumbing codes to be tested.
T	F	**14.**	An air test of the potable water supply and distribution piping is conducted in a manner similar to an air test on the sanitary drainage and vent piping.
T	F	**15.**	A final air test is a test of the plumbing fixtures and their connections to the sanitary drainage system.
T	F	**16.**	Since some plumbing fixture traps contain only a 2″ water seal, excessive pressure may blow out the traps and void the initial air test.
T	F	**17.**	If it is not possible to work alone in the building, turn off all flushometer valves and empty all water closet flush tanks during a final air test to prevent flushing of the fixtures.
T	F	**18.**	Piping must be exposed until it has been tested and inspected.

Multiple Choice

_____ **1.** The entire plumbing system must be tested and inspected to certify that systems meet ___ requirements.
    A. universal
    B. international
    C. plumbing code
    D. government

_____ **2.** Mechanical test plugs are available in ⅜ to___″ pipe sizes.
    A. 12
    B. 13
    C. 14
    D. 15

_____ **3.** Inflatable test plugs are available in ___″ pipe sizes.
    A. ¾ to 18
    B. ¾ to 19
    C. 1 to 18
    D. 1¾ to 21

_____ **4.** A(n) ___ plug is a test device installed on the inside of the pipe end.
    A. inspection
    B. pipe
    C. gauge
    D. inlet

_____ **5.** Air and hydrostatic tests are typically performed on plumbing systems by closing the appropriate piping and adding ___ air or water to pressurize the plumbing system.
   A. compressed
   B. cold
   C. hot
   D. humid

_____ **6.** The ___ of a mechanical test plug is inserted into the pipe or opening and the hex or wing nut is tightened.
   A. O-band
   B. O-ring
   C. P-ring
   D. P-band

_____ **7.** A(n) ___ cap is a reinforced rubber cap installed on the outside of an opening and secured in position with a stainless steel hose clamp.
   A. inspection
   B. plug
   C. test
   D. outlet

_____ **8.** A(n) ___ cap is a test device installed on the outside of a pipe end.
   A. pipe
   B. assembly
   C. outside
   D. tee

_____ **9.** To perform a building sewer air test or hydrostatic test, ___ in the building sewer are plugged with test plugs or caps.
   A. front main cleanouts
   B. test tees
   C. underground wyes
   D. all of the above

_____ **10.** Since plastic pipe caps or plugs are ___ to ABS DWV pipe to seal drain or vent openings, pipe caps and plugs cannot be reused when testing other plumbing systems.
   A. scald-cemented
   B. solvent-cemented
   C. melted
   D. soldered

_____ **11.** ___ are required to perform a hydrostatic test of a plumbing system.
   A. Water hoses
   B. Test plugs or caps
   C. Pipe plugs or caps
   D. all of the above

_____ **12.** Test caps are available in ___″ pipe sizes.
     A. ½ to 8
     B. 1 to 9
     C. 1½ to 8
     D. 1½ to 9

_____ **13.** If a hydrostatic test is applied to the entire system, all openings in the drainage and vent piping must be tightly sealed, except the highest vent opening above the ___.
     A. basement
     B. water closet
     C. kitchen sink
     D. roof

_____ **14.** Since 1′ of head of water equals .434 psi of pressure, a 100′ head (ten-story building) would equal ___ psi of pressure.
     A. 43.0
     B. 43.4
     C. 44.0
     D. 44.4

_____ **15.** In residential or light commercial construction, air is released from the inflatable ___ test plug placed in the front main cleanout.
     A. rubber
     B. plastic
     C. hose
     D. resin

_____ **16.** Potable water supply and distribution piping is typically tested at 1½ times the working water pressure or ___ psi.
     A. 50
     B. 100
     C. 150
     D. 200

_____ **17.** The test duration for a potable water supply and distribution piping air test is ___ hr.
     A. 6 to 24
     B. 6 to 30
     C. 12 to 24
     D. 12 to 30

_____ **18.** A(n) ___ leak detector is commonly used to locate small leaks in the potable water supply and distribution piping system.
     A. ultrasonic
     B. distribution
     C. concentration
     D. probe

_____ **19.** The entire sanitary drainage and vent piping system or sections of the system undergoing the hydrostatic test must maintain a ___′ head of water pressure for 15 min without the addition of water.
　　A. 4
　　B. 6
　　C. 8
　　D. 10

_____ **20.** If an air test is performed over the course of more than one day, the test plug in the front ___ opening must be removed.
　　A. tee
　　B. main cleanout
　　C. stack
　　D. vent

_____ **21.** ___ can affect the system between the times when the initial tests and final air test are performed.
　　A. Careless backfilling
　　B. Careless compaction
　　C. Use of heavy equipment over underground piping
　　D. all of the above

Completion

_____ **1.** A pressure change or decrease in water level indicates a(n) ___ in the plumbing system.

_____ **2.** A plumbing system inspection is performed by a plumbing inspector or ___.

_____ **3.** A(n) ___ test plug is an inflatable rubber device inserted into the plumbing system to seal openings during an air or hydrostatic test.

_____ **4.** A test ___ is a test device used to measure pressure within waste and vent, water, gas, air, or other piping systems to ensure that the system is maintaining the proper pressure.

_____ **5.** For a hydrostatic test, the plumbing system must be filled with water to a point equal to a ___′ head.

_____ **6.** A hydrostatic test, or ___ test, is a plumbing system test in which pipe openings are sealed with plugs or caps and the pipe is filled with water to provide a specified amount of pressure to the plumbing system to determine the tightness of the system.

_____ **7.** A(n) ___ test plug is a test device inserted into the end of a pipe or other opening and secured in position by tightening a hex or wing nut.

_____ **8.** A(n) ___ test is performed on the water main and water service piping to ensure the main and piping are free from defects and are able to convey water under pressure to a building.

_____ **9.** A(n) ___ water solution is applied to the pipe joints to identify leaks in the system.

_____ **10.** Compressed air contains a large amount of stored energy that presents a safety hazard if the plumbing system fails during ___.

_____ **11.** To avoid the difficulty encountered when losing an inflatable test plug, an inflatable ___ plug may be used to plug the stack.

_____ **12.** Since ___ in potable water supply and distribution piping are usually small, they are difficult to locate with a soapy water solution.

_____ **13.** A(n) ___ air test is performed after all plumbing fixtures are set and their traps are filled with water.

_____ **14.** A(n) ___ is a clear U-shaped tube partially filled with water; it is used to measure pressure within a closed system.

_____ **15.** A(n) ___ test is performed by filling fixture traps with water, plugging the building drain at the front main cleanout, and introducing a thick, odorous smoke into the system.

_____ **16.** A(n) ___ is a device constructed of a short section of large-diameter pipe and two reducer fittings in which a smoke cartridge is placed.

_____ **17.** To perform a(n) ___ test, fixture traps are filled with water and the building drain is plugged at the front main cleanout.

_____ **18.** Plumbing tests involve subjecting plumbing systems to ___.

Matching

Ultrasonic Leak Detector

_____ **1.** Headphones

_____ **2.** Ultrasonic noise generator

_____ **3.** Amplifier

_____ **4.** Contact probe

_____ **5.** Hand probe

_____ **6.** Sound concentrator

Short Answer

1. Describe how a hydrostatic test is performed on the sanitary drainage and vent piping in a tall building.

_____

_____

_____

_____

2. List the procedure for performing a plumbing test.

_____

_____

_____

_____

_____

_____

_____

_____

3. Explain why hydrostatic tests should not be performed when the temperature falls below 32°F.

_____

_____

4. List the three primary differences between tests performed on the potable water supply and distribution piping and on the sanitary drainage and vent piping.

_____

_____

_____

5. Explain why an air test should not be used on plumbing systems constructed of plastic unless the air test procedure is clearly approved by the manufacturer.

_____

_____

_____

6. List five safety precautions that should be taken when working with a plumbing system that is under pressure.

_____

_____

_____

_____

_____

7. List the procedure for performing a final air test.

_____

_____

_____

_____

_____

_____

_____

_____

_____

Activities

Hydrostatic Test Tools and Equipment

1. Indicate on the drawing the pipes to be plugged when performing a hydrostatic test of the entire sanitary drainage and vent piping system.

2. List the tools and equipment necessary to perform the test.

_____

_____

_____

_____

_____

_____

_____

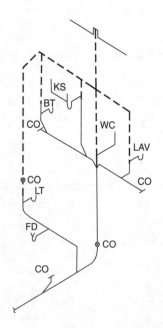

Air Test Tools and Equipment

1. Indicate on the drawing the pipes to be plugged when performing an air test of the entire sanitary drainage and vent piping system.

2. List the tools and equipment necessary to perform the test.

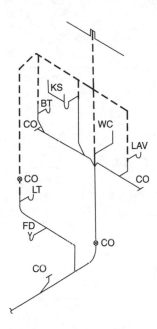

_____

_____

_____

_____

_____

_____

_____

_____

Final Air Test Procedure

1. The final air test of a plumbing system involves six steps, which are listed below. On the drawing, indicate where the step should be performed by writing the number of the step next to the proper trap, fixture, or opening.

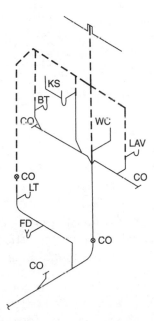

A. Fill fixture and floor drain traps with water
B. Plug roof stack openings
C. Plug building drain
D. Insert manometer hose through trap seal of water closet bowl
E. Fill manometer with water to 0″ mark on manometer
F. Attach hose between manometer and water closet
G. Insert air hose through trap seal and blow air into system

2. List the tools and equipment necessary to perform the final air test.

_____

_____

_____

Manometer Reading

1. Carefully examine the two diagrams of the manometers under test conditions. Explain what is happening in each situation.

_____

_____

_____

_____

_____

_____

_____

_____

_____

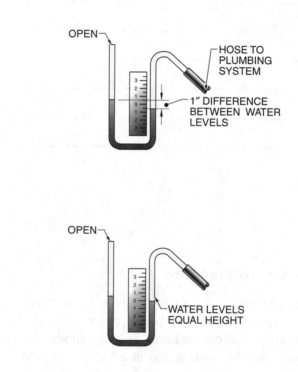

Projects

1. Obtain a set of residential prints and sketch an isometric piping drawing of the water supply and DWV system. Develop a plan for efficiently performing a hydrostatic test, air test, and final air test of the dwelling. Identify potential challenges that may exist when performing these tests.

2. Obtain a set of prints for a commercial building that includes a piping drawing. Determine whether the plumbing systems can be tested in their entirety or in sections. Identify the tools and equipment necessary to efficiently perform tests required in the local municipality.

Name_____ Date _____

True-False

T   F   **1.** If necessary, wood shims are installed under the bathtub to ensure that the fixture is level.

T   F   **2.** Rough-in sheets should be obtained each time a fixture or appliance is ordered since manufacturers frequently make design and dimension changes.

T   F   **3.** When a kitchen sink is located below a window, the waste and vent stack must be centered directly below the window.

T   F   **4.** The grade of the building sewer is calculated by dividing the run of the building sewer piping by the fall.

T   F   **5.** Prints should be studied before plumbing installation begins so a plumber can visualize the entire building and its structural details, all of which must be considered as the plumbing system is designed.

T   F   **6.** Upon approval of the DWV system for the building, all test plugs and appropriate test caps must remain in the piping to prevent impairing the flow of the system.

T   F   **7.** The water service is typically installed in a trench separate from the building sewer trench.

T   F   **8.** Prints for small buildings, such as one-family dwellings, provide information about fixtures and appliances and their locations, but do not typically include information about location of pipe and fittings and routing of the pipe throughout the building.

T   F   **9.** Dimensions for the fixtures and appliances and the plumbing equipment they require are always included on residential and light commercial prints.

T   F   **10.** The cold soft water piping starts at the outlet of the water heater.

T   F   **11.** Plumbing permits are typically required by municipalities for water service and building sewer connections, and for piping to be installed within a building.

T   F   **12.** Water supply installation for residential and light commercial buildings is typically performed in two phases—prepping and finishing.

T F **13.** Care must be taken not to permit heavy rocks or large pieces of dirt to fall directly on the building sewer and water service piping to avoid damaging them.

T F **14.** Hot water fixture supply piping is required to service each water closet.

Multiple Choice

_____ **1.** Wood backing, typically ___, is nailed or screwed between the studs to properly support wall-hung plumbing fixtures.
  A. 1 × 4s
  B. 2 × 6s
  C. 2 × 8s
  D. all of the above

_____ **2.** The water service is piped to a ___ box when the street mains are being installed.
  A. service
  B. start
  C. stop
  D. municipal plumbing

_____ **3.** When determining the building drain trench depth, the ___ must be known.
  A. fall of the building drain piping
  B. basement floor thickness
  C. building drain diameter
  D. all of the above

_____ **4.** Plumbing permits are usually obtained from the ___ by a plumbing contractor.
  A. municipal building department
  B. local fire department
  C. local police department
  D. County Clerk's office

_____ **5.** When a building sewer grade of more than ¼″ is required, the sewer is graded at ¼″ per foot toward the sanitary sewer connection, and two ___″ bends are used to make the final connection to the sewer.
  A. $\frac{1}{16}$
  B. $\frac{1}{8}$
  C. $\frac{1}{4}$
  D. $\frac{1}{2}$

_____ **6.** Prints indicate ___ location.
  A. fixture
  B. appliance
  C. cabinetry
  D. all of the above

_____ **7.** Floor-set plumbing fixtures, except ___, are set after the rooms in which they are to be installed have been decorated and are ready to be occupied.
    A. lavatories
    B. bathtubs, shower bases, and shower stalls
    C. water closets
    D. all of the above

_____ **8.** The locations of the water service connection and the sanitary sewer connection are marked on the lot with ___.
    A. blue spray paint
    B. a flag
    C. a stake
    D. all of the above

_____ **9.** The ___ must be installed before the drywall and wallcoverings are installed.
    A. shower base
    B. bathtub
    C. shower and bathtub fittings
    D. all of the above

_____ **10.** Specifications describe the ___ for plumbing fixtures and appliances.
    A. quality
    B. manufacturer and color
    C. type of piping material required
    D. all of the above

_____ **11.** The height of the double sanitary tee is calculated by subtracting the ___ and tailpiece length from the laundry tray height and rounding to the nearest $\frac{1}{2}$" increment.
    A. laundry tray depth
    B. clothes washer height
    C. clothes washer depth
    D. laundry tray vent

_____ **12.** Horizontal drainage piping should not have a grade greater than ___" per foot.
    A. $\frac{1}{4}$
    B. $\frac{1}{2}$
    C. $\frac{3}{4}$
    D. 1

_____ **13.** ___ surround the vent pipes passing through the roof and make the area of the surrounding roof watertight.
    A. Vent pipe gaskets
    B. Roof jackets
    C. Roof sleeves
    D. Vent pipe flanges

Completion

_____ 1. Final installation of roof jackets is typically performed by ___.

_____ 2. The ___ is the elevation of the lowest part of the inside of a horizontal pipe.

_____ 3. Cold soft water is not supplied to the kitchen sink faucet since the ___ content of soft water may adversely affect individuals with certain health conditions.

_____ 4. Plumbing contractors must refer to fixture and appliance ___ sheets to gain knowledge of the fixture and appliance requirements for the building.

_____ 5. A(n) ___ base is mixed and properly applied to the rough floor below the shower base per manufacturer specifications.

_____ 6. A(n) ___ is a drawing that provides the dimensions needed to rough in the water supply and DWV piping for fixtures and appliances.

_____ 7. Hot water originates at the ___ and flows to fixtures and appliances connected to the hot water piping system.

_____ 8. When a contractor applies for a plumbing permit, ___ and specifications of the plumbing work commonly must be submitted for review by the plumbing inspector to ensure conformity to code.

_____ 9. After reviewing the prints and specifications, a(n) ___ drawing is sketched to allow a plumber to visualize the construction of the water supply system.

_____ 10. A(n) ___, or bill of materials, is a list of all pipe, fittings, and other materials required to construct a plumbing system.

_____ 11. The building drain outlet and ___ connection depths at the front lot line must be known to determine the grade of the building sewer.

_____ 12. The 3″ soil stack is located by placing a(n) ___ string at the center of the opening in the first-floor common bathroom wall and lowering it to the basement.

_____ 13. When testing is completed and the sanitary waste and vent piping is approved, the basement trenches must be properly ___.

_____ 14. A(n) ___ is a small copper plate mounted on a concrete pier; the plate indicates its height above sea level or a previously established grade.

_____ 15. ___ is the installation of water distribution piping and setting of plumbing fixtures.

_____ 16. The water supply system should be filled with ___ to test for leaks.

_____ 17. ___, or rough plumbing, is the installation of parts of a plumbing system that can be completed prior to the installation of the fixtures.

Short Answer

1. What is the purpose of a plumbing permit?

_____

_____

_____

2. Why might a two-tier trench be dug for the installation of a water service and building sewer?

_____

_____

_____

_____

3. What work is performed by a plumber when installing a roof jacket?

_____

_____

_____

_____

Matching

Identifying Fittings

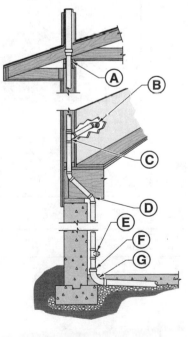

_____ **1.** 2 × 1½ coupling

_____ **2.** 1½″ test tee with plug

_____ **3.** 1½″ 90° elbow

_____ **4.** 2″ long-turn 90° elbow

_____ **5.** 1½″ sanitary tee

_____ **6.** 2 × 1½ bushing

_____ **7.** 1½″ 45° elbow

Rough-In Sheets

_____ **1.** Model number

_____ **2.** Rough-in drawing with dimensions

_____ **3.** Fixture description

_____ **4.** Fixture specification

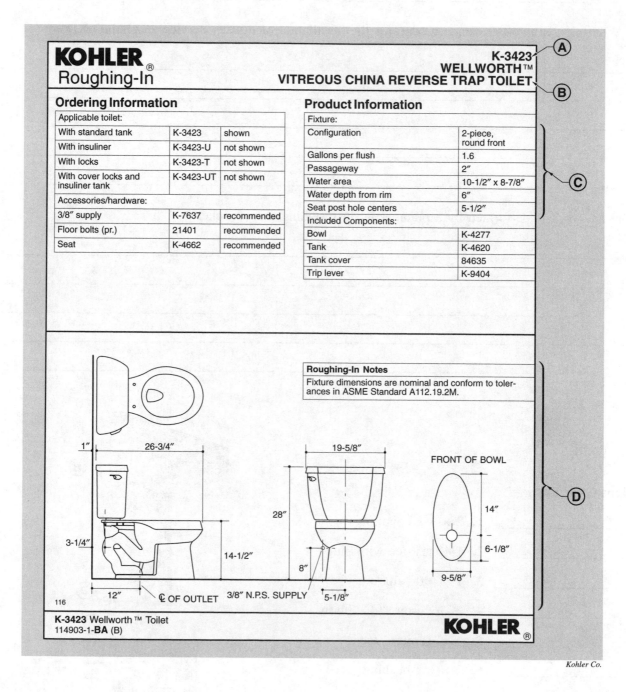

Reading Specifications

_____ 1. ___ water closet(s) is(are) required.

_____ 2. The color of the stall shower is ___.

_____ 3. The make of the kitchen sink is ___.

_____ 4. The water heater storage tank is ___ lined.

_____ 5. The capacity of the water heater is ___ gal.

_____ 6. The kitchen sink dimensions are ___.

_____ 7. The manufacturer's fixture model number for the bathtub is ___.

_____ 8. Footing drains are connected to the ___.

_____ 9. The heating capacity of the water heater is ___ gph.

_____ 10. The make and model of the water heater is ___.

Plumbing

Fixture	Number	Location	Make	MFR's Fixture Model No.	Size	Color
Sink	1	Kitchen	Elkay	PSR 3322	33 x 22	
Lavatory	1	Full Bathroom	Kohler	K-2210	17 x 14	White
Water closet	2	Bathroom	Kohler	K-3423		white
Bathtub	1	Full Bathroom	Kohler	K-506		
Shower over tub	1	Full Bathroom	Kohler	K-1513E		
Stall shower	1	¾ Bathroom	Mustee base	3248 M	32 x 48	Bone
Laundry trays	1	Basement	Mustee	26 W	24 x 40	White
Lavatory	1	¾ Bathroom	Kohler	K-2293-4		white
Shower valve	1	¾ Bathroom	Kohler	K-15141		
Lavatory faucet	2	Bathroom	Kohler	K-15182		

Bathroom accessories ☐ Recessed material _____ number _____ ☐ Attached material _____ number _____

Additional information

☒ Curtain rod ☐ Door ☐ Shower pan material _____   *(Show and describe individual system in complete detail in separate drawings and specifications according to requirements.)

Water supply ☒ public ☐ community system ☐ individual (private) system*

Sewage disposal ☒ public ☐ community system ☐ individual (private) system*

House drain (inside) ☐ cast iron ☐ tile ☒ other Plastic (ABS) House sewer (outside) ☐ cast iron ☐ tile ☒ other Plastic (ABS)

Water piping ☐ galvanized steel ☒ copper tube ☐ other   Sillcocks, number 2

Domestic water heater type Nat. Gas make and model A.O. Smith FSGL heating capacity 40.9 gph. 100° rise.

Storage tank material Glass lined capacity 40 gallons

Gas service ☒ utility company ☐ liq. pet. gas ☐ other _____ ☐ Gas piping ☐ cooking ☐ house heating

Footing drains connected to ☐ storm sewer ☒ sanitary sewer ☐ dry well ☐ Sump pump make and model _____

capacity _____ discharges into _____

Activities

Sketching a DWV System

1. Using the plan view of the bathroom and kitchen sink, sketch an isometric piping drawing of the DWV system.

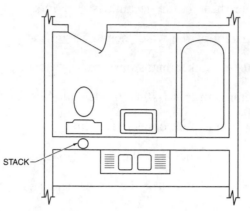

Sketching and Sizing a DWV System

1. Using the plan view of the bathroom and kitchen sink, sketch an isometric piping drawing with stack group venting used to vent the fixtures. Size the drawing according to local code.

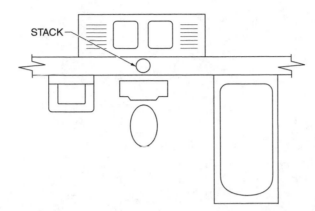

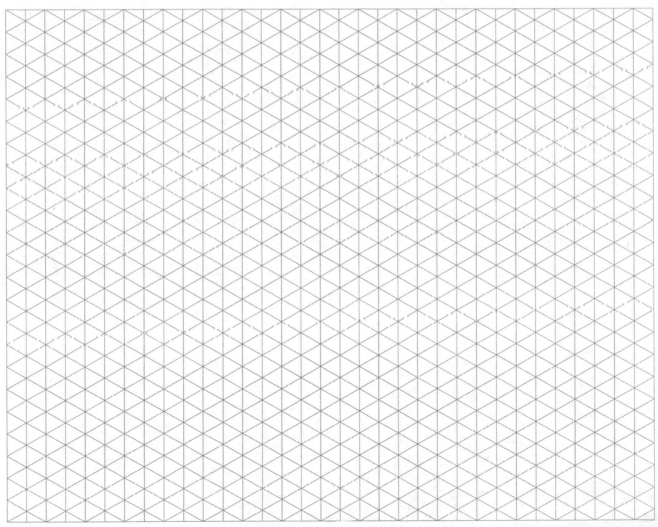

DWV System Design

1. Use the set of residential prints to create an isometric piping drawing of the DWV system. A soil stack, located in each wall behind the water closets, is indicated on the prints. Size the prints based on the local plumbing code.

DWV Material List

1. Use the isometric piping drawing from the DWV System Design activity to create a DWV material list. Include all pipe and fittings, fixtures, appliances, and trim. Refer to manufacturer catalogs for specifications.

DMV MATERIAL LIST		
JOB NAME	**DATE**	
	PREPARED BY	
QTY	**SIZE**	**DESCRIPTION OF MATERIAL**

Water Supply System Design

1. Use the set of residential prints to create an isometric piping drawing of the hot and cold water distribution system. A water service and water meter are indicated on the prints. Size the hot and cold water distribution system based on the local plumbing code.

Water Supply Material List

1. Use the isometric piping drawing from the Water Supply System Design activity to create a water supply material list. Include all pipe and fittings, fixtures, appliances, and trim. Refer to manufacturer catalogs for specifications.

WATER SUPPLY MATERIAL LIST		
JOB NAME	**DATE**	
	PREPARED BY	
QTY	**SIZE**	**DESCRIPTION OF MATERIAL**

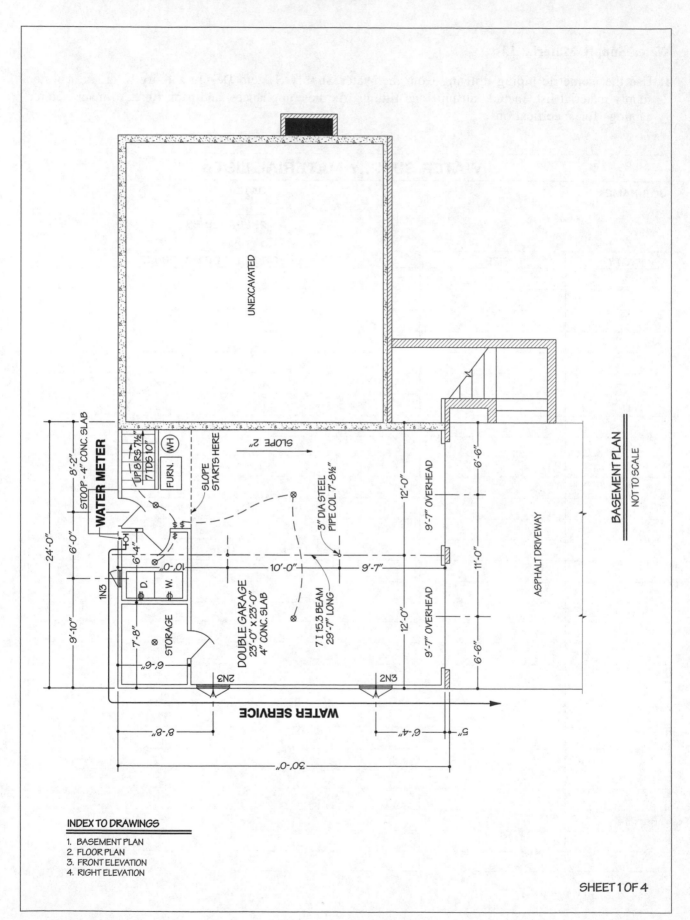

WATER METER

UNEXCAVATED

STOOP - 4" CONC. SLAB

8'-2"

WH

UP 8 RS 7½"
7 TDS 10"

FURN.

SLOPE STARTS HERE

SLOPE 2"

3" DIA STEEL
PIPE COL. 7'-8½"

24'-0"

6'-0"

9'-10"

1N3

1'-0"

6'-4"

10'-0"

10'-0"

9'-7"

7 I 15.3 BEAM
29'-7" LONG

12'-0"

12'-0"

9'-7" OVERHEAD

9'-7" OVERHEAD

6'-6"

6'-6"

11'-0"

ASPHALT DRIVEWAY

D.   W.

7'-8"

STORAGE

6'-6"

DOUBLE GARAGE
23'-0" x 23'-0"
4" CONC. SLAB

2N3

2N3

WATER SERVICE

8'-8"

6'-4"

5"

30'-0"

BASEMENT PLAN
NOT TO SCALE

INDEX TO DRAWINGS

1. BASEMENT PLAN
2. FLOOR PLAN
3. FRONT ELEVATION
4. RIGHT ELEVATION

SHEET 1 OF 4

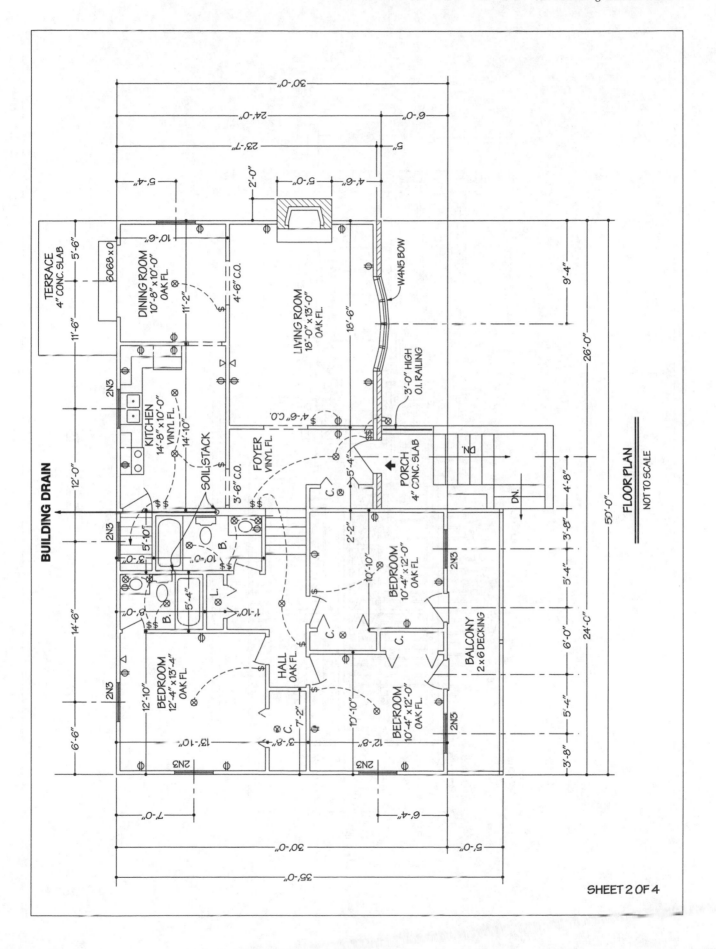

BUILDING DRAIN

TERRACE
4" CONC. SLAB

6068 x 0

DINING ROOM
10'-8" x 10'-0"
OAK FL.

4'-6" C.O.

LIVING ROOM
18'-0" x 13'-0"
OAK FL.

W4N5 BOW

3'-0" HIGH
O.I. RAILING

KITCHEN
14'-8" x 10'-0"
VINYL FL.

SOIL STACK

4'-6" C.O.

FOYER
VINYL FL.

3'-6" C.O.

PORCH
4" CONC. SLAB

DN.

DN.

B.

L.

B.

BEDROOM
12'-4" x 13'-4"
OAK FL.

HALL
OAK FL.

C.

C.

C.

BEDROOM
10'-4" x 12'-0"
OAK FL.

BALCONY
2 x 6 DECKING

C.

BEDROOM
10'-4" x 12'-0"
OAK FL.

FLOOR PLAN
NOT TO SCALE

SHEET 2 OF 4

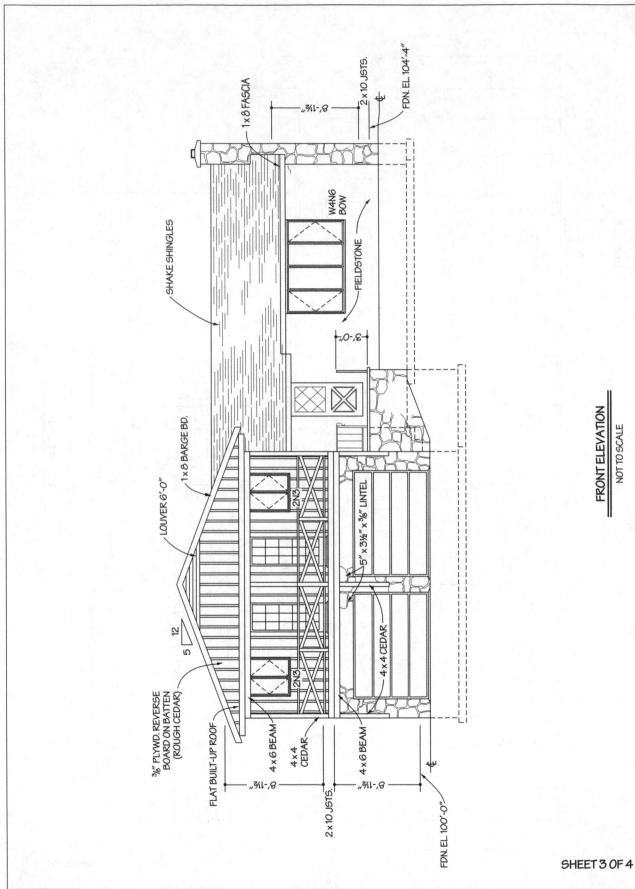

FRONT ELEVATION
NOT TO SCALE

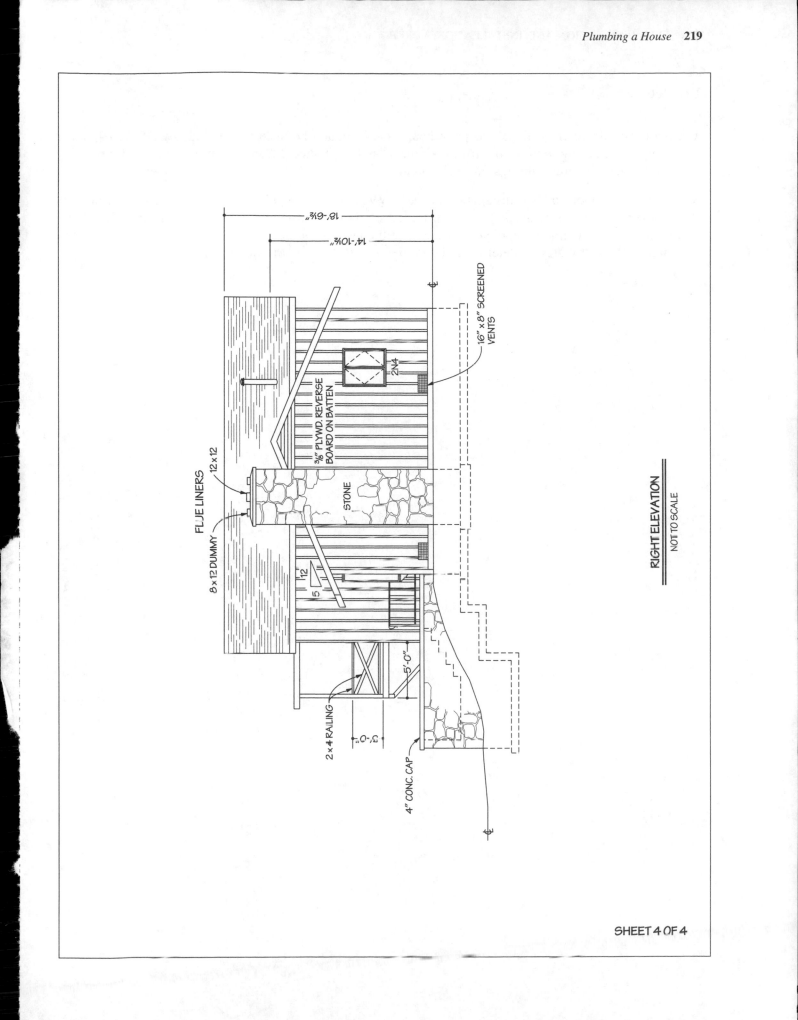

RIGHT ELEVATION
NOT TO SCALE

Projects

1. Obtain manufacturer catalogs and price books and estimate the material cost for the DWV and water supply systems, including all fittings, fixtures, and appliances. Compare prices and features using several manufacturer catalogs and price books.

2. Estimate the labor cost for installation of the DWV and water supply systems from project 1. Sources for labor rate information include local trade associations, printed labor tables, and historical labor rate data. Ask a plumbing contractor to estimate the installation of the DWV and water supply systems. Compare the estimates to determine similarities and differences in the estimates.